Synthesis Lectures on Operations Research and Applications

This series focuses on the use of advanced analytics in both industry and scientific research to advance the quality of decisions and processes. Written by international experts, modern applications and methodologies are utilized to help researchers and students alike to improve their use of analytics. Classical and cutting-edge topics are presented and explored with a focus on utilization and application across a range in practical situations.

John Hooker

Logic-Based Benders Decomposition

Theory and Applications

 Springer

John Hooker
Carnegie Mellon University
Pittsburgh, PA, USA

ISSN 2770-6303   ISSN 2770-6311   (electronic)
Synthesis Lectures on Operations Research and Applications
ISBN 978-3-031-45038-9   ISBN 978-3-031-45039-6   (eBook)
https://doi.org/10.1007/978-3-031-45039-6

This Springer imprint is published by the registered company Springer Nature Switzerland AG
The registered company address is: Gewerbestrasse 11, 6330 Cham, Switzerland

Paper in this product is recyclable.

Preface

Benders decomposition is a well-known and often very effective tool for solving hard optimization problems. It is especially advantageous when a problem reduces to a much simpler subproblem after certain variables are fixed. It benefits from an ingenious learning mechanism based on Benders cuts.

Despite its success, the classical Benders method has limited applicability, because the subproblem must be a linear programming problem. Fortunately, its underlying problem-solving strategy is much more general than may be evident at first. Benders cuts can be viewed as arising from logical inference rather than the specific properties of linear programming. This enables a substantial generalization to *logic-based Benders decomposition* (LBBD), in which the subproblem can, in principle, be any optimization problem. This, in turn, opens the door to a much broader range of applications.

This book is intended as a comprehensive guide to the LBBD user. It presents a unified account of LBBD theory as it has developed over the last two decades. It provides an in-depth tutorial on how to develop effective logic-based cuts for a given problem. It explains such related ideas as branch-and-check methods and combinatorial Benders cuts, the latter being a special case of logic-based cuts. It offers practical suggestions for crafting a successful LBBD implementation for a given application.

LBBD must, in fact, frequently be tailored to the specific application to harness its full potential. With this in mind, nearly half the book is devoted to a compendium of 147 different LBBD applications, ranging from transportation and supply chain management to hospital scheduling and disaster preparation. It describes how 226 published articles adapt the LBBD framework to these problems. This repository of domain-specific solutions not only demonstrates LBBD's wide applicability, but it can serve as a source of ideas for addressing the problem at hand.

Pittsburgh, USA   John Hooker
June 2023

v

Contents

Introduction

1.1 Generalizing Benders Decomposition

Benders decomposition, introduced in 1962 by Jacques Benders [20], is one of the best known and most successful strategies for solving hard optimization problems. It is designed for problems that become easier when certain variables are assigned fixed values, thus creating a more tractable *subproblem*. It solves a problem by enumerating various assignments to the fixed variables and solving the subproblems that result, with the aim of identifying an optimal solution.

While Benders decomposition has many successful applications, it is restricted by the fact that the subproblem must be a linear programming problem—or a convex nonlinear programming problem in Geoffrion's 1972 extension of the method [103]. There are a wide range of potential applications in which the subproblem simplifies without yielding a linear or nonlinear programming problem, often by decoupling into smaller problems. Benders decomposition in its classical form is not suitable for applications of this sort. Nonetheless, its underlying problem-solving idea is much more general than may be evident at first, and it can be extended to a substantially broader class of problems.

Logic-based Benders decomposition (LBBD), which dates from the 1990s, carries out this extension by allowing the subproblem to be *any* optimization problem, at least in principle. The method is "logic-based" in the sense that logical inference plays a key role in its conception, and not because there is any need for the problem statement to consist of logical formulas or have any other relation to formal logic. The enhanced generality of LBBD has enabled its application to a greatly expanded range of real-world problems. This can result in reductions of solution time of several orders of magnitude relative to the previous state of the art.

This book is intended to show how LBBD can be applied to a wide range of optimization problems. It begins by laying out the basic theory of LBBD, along with some variations

© The Author(s), under exclusive license to Springer Nature Switzerland AG 2024
J. Hooker, *Logic-Based Benders Decomposition*, Synthesis Lectures on Operations Research and Applications, https://doi.org/10.1007/978-3-031-45039-6_1

and enhancements. Because LBBD can, and often must, be tailored to fit the structure of a given application to obtain fast convergence, the book proceeds to illustrate how this may be accomplished for a variety of problem structures.

1.2   The Fundamental Idea

A Benders method benefits from the fact that subproblems can be solved relatively quickly as it searches for an optimal solution. Yet its primary problem-solving power derives from *Benders cuts* that guide the search. These cuts are constraints that bound the quality of solutions that would result from fixed variable assignments other than those already tried, based on an analysis of previous subproblem solutions. The Benders cuts are accumulated in a *master problem* that is solved to identify the next set of assignments and thereby the next subproblem to solve. The process continues until the optimal value of the master problem, and the best optimal subproblem value obtained so far, converge to the same value. This normally occurs after only a small fraction of the solution space has been explored.

Classical Benders decomposition obtains cuts by solving the linear programming dual of the subproblem. LBBD extends the classical method by observing that the linear programming dual is a special case of an *inference dual* that can be defined for any optimization problem, thus placing no restriction on the form of the subproblem. The inference dual seeks the best bound on a problem's optimal value that can be *logically deduced* from its constraint set. The brilliant maneuver that underlies Benders decomposition is to ask what bound *this same proof* can deduce from the subproblem constraints if the master problem variables are fixed to different values. In the classical Benders method, the proof is encoded as a set of dual multipliers that give rise to a Benders cut in the form of a linear inequality constraint. Logical inference therefore lies at the heart of classical Benders decomposition, and once this is recognized, an extension to LBBD becomes possible. One need only replace linear programming duality with inference duality and write a Benders cut that is based on logical inference.

Formulating a classical Benders cut is straightforward because it always follows the same pattern. By contrast, constructing other types of logic-based cuts typically requires a separate analysis (and often some ingenuity) for every problem class. This can be viewed as a weakness of LBBD, but it also provides an opportunity to design cuts that exploit a problem's special structure. Indeed, one might argue that solution of hard combinatorial problems typically requires problem-specific methods. While we are beginning to see general-purpose LBBD solvers that rely on generic logic-based cuts, we can nonetheless expect many problem classes to require or benefit from hand-crafted cuts.

A major objective of this book is to assist cut design by mining the rapidly growing LBBD literature for ideas on how to exploit the mathematical structure of a particular application. There is a large body of experience to draw from, due to the remarkable number and variety of existing applications, covering such diverse areas as supply chain logistics, computer processor scheduling, organ transplantation, wind turbine maintenance, search-and-rescue operations, and many others.

1.3   Early Developments

The logic-based approach of LBBD was foreshadowed in a 1990 paper of Jeroslow and Wang [138], who showed that the linear programming dual of a Horn clause system can be viewed as an inference problem. A Horn system is a set of specially structured logical propositions whose satisfiability can be checked by linear programming, as well as by an inference method known as unit resolution. When the system is unsatisfiable, a solution of the classical linear programming dual can be read directly from the structure of a unit resolution proof of infeasibility. The key implication of this work for future developments is that the dual problem can be treated as a logical inference problem.

Horn clauses in fact comprise the subproblem in what might be interpreted, in retrospect, as the first clear application of LBBD. It is described in a 1995 paper of Hooker and Yan [127], who solve logic circuit verification problems using a specialized Benders method in which the inputs to the circuit are master problem variables, and Horn clauses represent the circuit design. The first explicit description of LBBD as a general method appeared in the year 2000 [122]. Early computational experiments (for machine assignment and scheduling) are reported in a 2001 paper of Jain and Grossmann [136], and LBBD was further developed and tested computationally in a 2003 paper of Hooker and Ottosson [126].

Two additional developments occurred during the early years. One is *branch and check*, a variation of LBBD that can be used when the master problem is a mixed integer/linear programming (MILP) problem solved by a branching method. Branch and check solves the master problem only once, rather than repeatedly as in standard LBBD. Integer solutions encountered during the branching process are sent to the subproblem to obtain logic-based Benders cuts on the fly. These cuts are used alongside traditional cutting planes while solving the master problem. They differ from the cutting planes in a traditional branch-and-cut method, however, partly because they are valid only when one takes into account the subproblem constraints. Branch and check was initially described in Section 19.6 of [122] and first tested computationally in 2001 by Thorsteinsson [244], who coined the term "branch and check."

A second development is that of *combinatorial Benders cuts*, introduced in 2006 by Codato and Fischetti [60]. These are a type of logic-based cut used in a specialized version of branch and check. Codato and Fischetti use the cuts to solve MILP problems with "big-M" constraints, which are very common but notorious for making problems hard to solve. Their article shows how to get rid of the big-M constraints by accounting for them in the way a linear programming subproblem is constructed, thus substantially accelerating solution. The term "combinatorial Benders cut" is sometimes used to denote logic-based Benders cuts in general, but we reserve the term for logic-based cuts derived from a linear programming subproblem, as this is how it seems most often to be understood.

Surveys of the Benders decomposition literature can be found in [125, 210], the latter dealing specifically with LBBD.

1.4   A Motivating Example

One of the earliest and most frequent applications of LBBD is to assignment and scheduling problems, in which the master problem assigns tasks to facilities or agents, and the subproblem schedules the tasks assigned to each agent. A small example of this kind will illustrate the basic ideas of LBBD.

We have four jobs, any of which can be processed in shop 1 or shop 2. Each job j has a processing time p_{ij} in shop i, and it must be processed within the time window $[r_j, d_j]$. We wish to assign the jobs to shops and schedule them so as to minimize makespan, which the finish time of the last job to finish. The specific problem data appear in Table 1.1. Note that shop 1 processes jobs 1 and 3 more slowly than shop 2, while jobs 2 and 4 have the same processing time in either shop.

To write an optimization model for a problem of this form, suppose there are n jobs, and let the binary variable $x_{ij} = 1$ when job j is assigned to shop i. Then if we let s_j denote the start time of job j, the problem is

$$
\min_{M,x,s} \left\{ M \left| \begin{array}{l} M \geqslant s_j + \sum_i p_{ij} x_{ij}, \; \sum_i x_{ij} = 1, \; s_j \geqslant r_j, \; \text{all } j \\[2mm] s_j + \sum_i p_{ij} x_{ij} \leqslant s_k \; \text{ or } \; s_k + \sum_i p_{ik} x_{ik} \leqslant s_j, \\[2mm] \qquad\qquad\qquad\qquad \text{all } j, k \text{ with } j < k \text{ and } x_{ij} = x_{ik} = 1, \; \text{all } i \\[2mm] x \in S; \; x_{ij} \in \{0,1\}, \; \text{all } i, j \end{array} \right. \right\}
$$

where $s = (s_1, ..., s_n)$, and x is the matrix of variables x_{ij}. The first line of constraints define the makespan M, ensure that every job is assigned to exactly one agent, and require that jobs start no earlier than their release time. The second line prevents jobs from overlapping

Table 1.1 Data for a small example problem

Job j	r_j	d_j	p_{1j}	p_{2j}
1	3	6	3	2
2	3	5	1	1
3	0	5	3	2
4	3	6	1	1

Table 1.2 Benders iterations for a small example problem

Iteration	Master problem optimal value	Job assignments $(x_{11}, \ldots, x_{14})$	Subproblem optimal value[†]	Subproblem solution $(s_1, \ldots, s_4)$
1		$(1, 1, 0, 0)$	∞	
2	0	$(0, 1, 0, 1)*$	5	$(3, 4, 0, 3)*$
3	4	$(1, 0, 0, 0)$	6	$(3, 3, 0, 4)$
4	5	$(0, 0, 1, 1)$		

*optimal solution
[†] ∞ indicates infeasible subproblem

by requiring, for every pair of jobs in the same shop, that one finish before the other starts. The constraint $x \in S$ covers any restrictions on the assignments.

The problem decomposes naturally into an assignment problem and a scheduling problem. This is particularly advantageous because the scheduling subproblem decouples into a separate problem for each shop once assignments have been made. Thus for a particular assignment $\bar{x}$, we have the following scheduling problem for each shop i:

$$\min_{M_i, s} \left\{ M_i \;\middle|\; \begin{array}{l} M_i \geqslant s_j + p_{ij}, \quad s_j \geqslant r_j, \quad \text{all } j \in J_i \\ s_j + p_{ij} \leqslant s_k \text{ or } s_k + p_{ik} \leqslant s_j, \quad \text{all } j, k \in J_i \text{ with } j < k \end{array} \right\}$$

where $J_i = \{j \mid \bar{x}_{ij} = 1\}$ is the set of jobs assigned to shop i. If M_i is the makespan in shop i, then $\max_i \{M_i\}$ is the overall makespan we wish to minimize.

We now proceed to solve the problem by LBBD, which enumerates assignments to the variables x_{ij} and observes the smallest makespan that results. The progress of the procedure is summarized in Table 1.2.

Iteration 1. Suppose we begin by assigning jobs 1 and 2 to shop 1, and jobs 3 and 4 to shop 2, so that $(x_{11}, x_{12}, x_{13}, x_{14}) = (1, 1, 0, 0)$, and $(x_{21}, x_{22}, x_{23}, x_{24})$ is the complement $(0, 0, 1, 1)$. The resulting minimum makespan problem for each shop is illustrated by Gantt charts in Fig. 1.1. The horizontal dimension is time, and the brackets indicate time windows. The processing time is shown by a heavy line within each bracket.

It is clear from Fig. 1.1a that assigning jobs 1 and 2 to shop 1 creates a scheduling problem with no feasible solution. We therefore write a simple *nogood cut* indicating that this assignment must be avoided:

$$x_{11} + x_{12} \leqslant 1 \tag{1.1}$$

This cut actually excludes four assignments of jobs to shops, since there are four assignments in which shop 1 receives jobs 1 and 2 (possibly among others). All of these assignments are infeasible, because assigning jobs 1 and 2 already creates infeasibility.

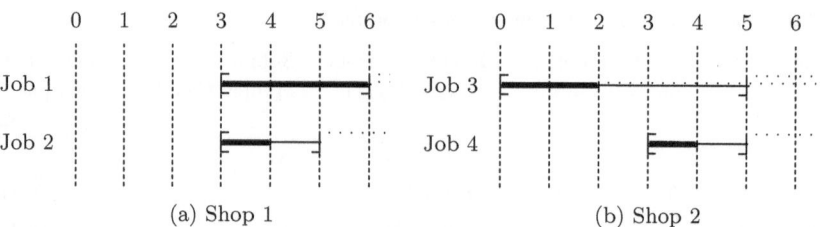

Fig. 1.1 Scheduling subproblems for shops 1 and 2 in the first iteration of LBBD applied to a small assignment and scheduling problem. The horizontal dimension is tine, brackets indicate tine windows, and heavy lines indicate processing tine

Figure 1.1b illustrates an optimal solution for shop 2, which received jobs 3 and 4. Since the minimum makespan is 4, we can write a valid cut

$$M \geqslant 4(x_{23} + x_{24} - 1)$$

where M represents the makespan of the original problem. The cut indicates that if jobs 3 and 4 are both assigned to the shop, the overall makespan must be at least 4, whereas if one or both jobs are assigned elsewhere, no meaningful bound is imposed.

However, there may be an opportunity to strengthen the cut. Perhaps the same minimum makespan would result if only one of the jobs were assigned. In principle, we could examine the proof of optimality generated by the scheduler (i.e., the solution of the inference dual) to determine which job assignments are premises of the proof. Yet since solvers typically do not provide this information, we might tease out part of the proof structure by rescheduling the shop after removing each job, one at a time, from its assignment. We find that the minimum makespan remains 4 after removing job 3. Thus job 3 is inessential to the optimality proof, and we can write the stronger cut

$$M \geqslant 4x_{24} \qquad (1.2)$$

The cuts says that if we wish to obtain a smaller makespan than 4, we must avoid assigning job 4 to this shop. A similar strategy can be used in an attempt to strengthen a feasibility cut like (1.1), but we find that removing either job 1 or job 2 relieves the infeasibility, which means this particular cut cannot be strengthened.

The strengthened cut (1.2) is useful but has nothing to say if only job 3 is assigned. We can obtain a meaningful bound in this case by applying an *analytical* Benders cut. This type of cut infers additional bounds based on the minimum makespan solution for the current assignment and an analysis of the problem structure. We will show in Chap. 3 that an analytical cut consisting of the following inequalities is valid for a makespan problem in shop i:

$$M \geqslant M_i^* - \sum_{j \in J_i}(1 - x_{ij})\big(p_{ij} + \max\{0, r_j - r_{\min} - p_{\min}\}\big) - (d_{\max} - d_{\min})$$
$$M \geqslant M_i^* - \sum_{j \in J_i}(1 - x_{ij})\big(p_{ij} + \max\{0, r_j - r_{\min} - p_{\min}\} + (d_{\max} - d_{\min})\big) \tag{1.3}$$

where M_i^* is the current minimum makespan for shop i, $r_{\min}$ is the earliest release time of jobs assigned to shop i, $d_{\min}$ and $d_{\max}$ the earliest and latest deadline, and $p_{\min}$ the smallest processing time of these jobs. The cut basically says that if there is a proof of makespan M_i^* for the jobs in J_i (the jobs currently assigned), there must be a proof of the bounds in (1.3) for subsets of these jobs. In the current instance, both inequalities in (1.3) evaluate to

$$M \geqslant 4 - 2(1 - x_{23}) - 3(1 - x_{24}) \tag{1.4}$$

Unlike the nogood cut, this cut provides a meaningful bound when only job 3 is assigned. The bound of 1 is quite weak in this case, but the analytical cuts are stronger in the next iteration.

Iteration 2. At this point we could simply try another assignment to the x_{ij} variables, but we prefer to let the cuts (1.1), (1.2) and (1.4) guide our search. We therefore formulate a *master problem* that minimizes makespan subject to these cuts and original constraints on the x_{ij}s:

$$\min_{M,x} \left\{ M \;\middle|\; \begin{array}{l} x_{11} + x_{12} \leqslant 1, \; M \geqslant 4x_{24}, \; M \geqslant 3 - 2(1 - x_{23}) - 3(1 - x_{24}) \\ x_{1j} + x_{2j} = 1, \; x_{1j}, x_{2j} \in \{0, 1\}, \; \text{all } j \end{array} \right\} \tag{1.5}$$

An optimal or near-optimal solution of this problem provides a reasonable suggestion of which assignment to the x_{ij}s to try next, because it results in a small makespan while taking into account what we have learned so far.

One optimal solution of (1.5) has value $M = 0$ and assigns jobs 2 and 4 to shop 1, with the remaining jobs assigned to shop 2. The scheduling problems that result are illustrated in Fig. 1.2. The minimum makespan solution for shop 1 shown in Fig. 1.2a has makespan 5 and immediately yields the cut

$$M \geqslant 5(x_{12} + x_{14} - 1) \tag{1.6}$$

The analytical cut (1.3) becomes

$$M \geqslant 5 - (1 - x_{12}) - (1 - x_{14}) \tag{1.7}$$

The minimum makespan solution for shop 2 in Fig. 1.2b yields a strengthened cut

$$M \geqslant 5x_{21} \tag{1.8}$$

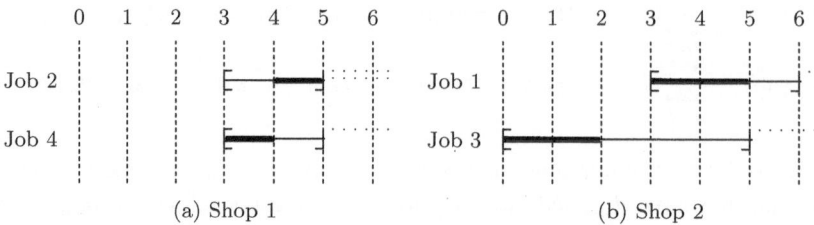

Fig. 1.2 Scheduling subproblems in the second iteration of LBBD

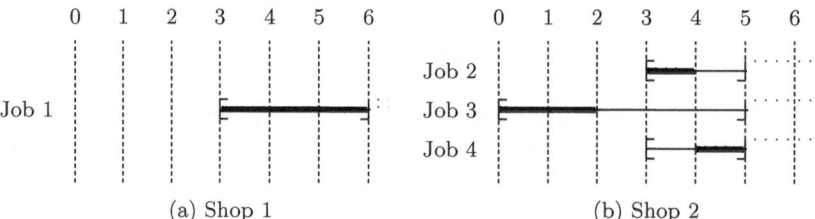

Fig. 1.3 Scheduling subproblems in the third iteration of LBBD

and analytical cut

$$M \geqslant 4 - 3(1 - x_{21}) - 2(1 - x_{23})$$
$$M \geqslant 5 - 4(1 - x_{21}) - 3(1 - x_{23})$$

$$(1.9)$$

Iteration 3. If we add cuts (1.6)–(1.9) to the master problem (1.5), its optimal value is 4. One of the optimal solutions is $(x_{11}, x_{12}, x_{13}, x_{14}) = (1, 0, 0, 0)$, which assigns job 1 to shop 1 and the other jobs to shop 2. The optimal makespan is 6 in shop 1 and 5 in shop 2, for an overall makespan of 6 (Fig. 1.3). Shop 1 generates the nogood cut $M \geqslant 6x_{11}$. Shop 2 also generates cuts, but as it happens, adding $M \geqslant 6x_{11}$ to the master problem will lead to immediate convergence.

Iteration 4. The optimal value of the updated master problem is 5, which is a lower bound on the minimum makespan, because the master problem is a relaxation of the original problem. But the best makespan found so far is also 5, which is an upper bound, because the subproblem is a restriction of the original problem. Thus 5 must be the optimal makepan. The corresponding solution

$$(x_{11}, x_{12}, x_{13}, x_{14}) = (0, 1, 0, 1), \quad (s_1, s_2, s_3, s_4) = (3, 4, 0, 3)$$

is therefore optimal. If there had been a gap between the master problem value and best makespan, we would have continued solving subproblems and generating cuts until the gap disappears, or the master problem becomes infeasible (indicating infeasibility of the original problem).

We note that if the optimal solution $\bar{x} = (0, 0, 1, 1)$ of the last master problem is sent to the subproblem, the resulting minimum makespan is 6, not 5. This cautions us that the solution values of the master problem variables at termination need not be optimal for the original problem. Rather, one must use the master problem solution in an iteration in which the subproblem delivers the optimal value of the original problem. In the present example, this occurs in iteration 2.

LBBD offers two clear advantages for a problem of this sort. One is that it allows one to apply a Benders strategy and take advantage of the separability of the scheduling problems, even though the subproblem is a combinatorial problem rather than the linear or nonlinear programming problem required by classical Benders methods. A second advantage is that the assignment and scheduling aspects of the problem can be addressed by methods appropriate to each. The assignment task in the master problem is well suited for mixed integer/linear programming, and the scheduling task in the subproblem to constraint programming or a specialized scheduling method [27].

1.5   Plan of the Book

About half of the book is devoted to theory, and the remainder to applications. The theory section consists of Chaps. 2, 3 and 4. Chapter 2 presents the basic concepts, beginning with inference duality and proceeding to develop the LBBD algorithm and prove finite convergence. Its shows how classical Benders decomposition is a special case of LBBD and suggests some alternative perspectives on LBBD that can provide additional insight into the method. It then takes up the important special case in which the subproblem decouples into smaller problems, and concludes with a number of practical suggestions for improving the performance of LBBD.

Chapter 3 focuses on the all-important matter of designing effective logic-based cuts. It develops a taxonomy of cuts, which are broadly divided into nogood cuts, analytical cuts, and explanation-based cuts. Nogood cuts are based on the intelligent sampling of solutions of the subproblem and can be strengthened in various ways described in the chapter. Analytical cuts are derived from a structural analysis of the subproblem. As object lessons in the craft of cut design, the chapter shows in some depth how strong analytical cuts can be obtained for scheduling, vehicle routing, and packing problems, as well as presenting a class of generic analytical cuts. It then illustrates how explanation-based cuts can be obtained directly from an inference dual solution.

Chapter 4 covers variations and special cases of LBBD. Variations include branch-and-check and enumerative methods, multilevel LBBD, and dynamic variable partitioning. Special cases include combinatorial Benders cuts for MILP, and stochastic and robust LBBD. The chapter also discusses how the master problem can be enhanced with a subproblem relaxation. It concludes with a brief look at how LBBD might be automated in a general-purpose solver, along with the pros and cons of doing so.

The applications-oriented portion of the book makes an important contribution to theory as well, by showing how logic-based cuts can exploit specific problem structure. Contained entirely in a long Chap. 5, it surveys a wide range of published applications and describes briefly how the LBBD framework was adapted to these problems and what kind of logic-based cuts were used. The applications are classified as dealing with transportation, manufacturing and production, supply chains and logistics, computing and telecommunications, health and medicine, disaster management, and miscellaneous scheduling and other problems, as well as such abstract problem classes as linear complementarity and operator counts for automated planning. Most of these are further divided into subcategories. The primary aim of this chapter is to allow readers to mine the ingenious solutions found by previous investigators as a possible source of ideas for the problem at hand.

Basic Theory

2

2.1 Introduction

The theory of logic-based Benders decomposition relies fundamentally on inference duality, and we begin the chapter with an exposition of this concept. We then state the basic LBBD algorithm, both in general and when the subproblem is a feasiblity problem. We show that if the master problem variables have finite domains, which is the norm in practice, the algorithm must converge finitely to an optimal solution or proof of infeasiblity.

Classical Benders decomposition is perhaps best understood as a special case of LBBD, because LBBD lays bare the essence of its problem-solving strategy. It is therefore presented here as an instance of LBBD in which the inference dual is the linear programming dual.

Some alternate perspectives on LBBD can yield insight into the method. For example, LBBD can be seen as partially computing the projection of the problem's feasible set (or more precisely, the problem's epigraph) onto the master problem variables. It is also interpretable as an exact form of large neighborhood search, particularly if one allows for dynamic variable partitioning from one iteration to the next. The most popular algorithms in highly-developed propositional satisfiability (SAT) solvers and SAT-modulo-theory solvers can also be viewed as forms of LBBD with dynamic variable partitioning.

The chapter next takes up the important case in which the subproblem decouples into smaller problems, with the primary aim of clarifying when and why Benders cuts generated for the smaller problems are valid for the entire subproblem. It concludes with some practical suggestions for building efficient LBBD implementations. A glossary of notation used throughout the book is given in Table 2.1.

© The Author(s), under exclusive license to Springer Nature Switzerland AG 2024
J. Hooker, *Logic-Based Benders Decomposition*, Synthesis Lectures on Operations
Research and Applications, https://doi.org/10.1007/978-3-031-45039-6_2

Table 2.1 Notation used throughout the book

$x = (x_1, \ldots, x_n)$	Master problem variables
$y = (y_1, \ldots, y_m)$	Subproblem variables
w	Auxiliary variables in the master problem
$\bar{x}$	Current optimal solution of the master problem
x^k	Optimal solution of the master problem in iteration k
$\text{SP}(\bar{x})$	Subproblem that results from master problem solution $\bar{x}$
$v^*(\bar{x})$	Optimal value of subproblem $\text{SP}(\bar{x})$
$v^*(J)$	$v^*(\bar{x})$, when $\bar{x}_j = 1$ for $j \in J$ and $\bar{x}_j = 0$ for $j \notin J$
$z \geq B_{\bar{x}}(x)$	An optimality Benders cut derived from subproblem $\text{SP}(\bar{x})$
$F_{\bar{x}}(x)$	A feasibility Benders cut derived from subproblem $\text{SP}(\bar{x})$
$J_1(x)$	$\{ j \in J \mid x_j = 1 \}$
$J_0(x)$	$\{ j \in j \mid x_j = 0 \}$

2.2   The Inference Dual

Consider a general optimization problem of the form

$$\min_x \left\{ f(x) \mid C(x), \ x \in \mathcal{D}_x \right\} \tag{2.1}$$

where $C(x)$ represents a constraint set containing variables in $x = (x_1, \ldots, x_n)$, and $\mathcal{D}_x$ is the domain of x (for example, tuples of nonnegative reals or integers). We refer to (2.1) as the *primal* problem. The *inference dual* of (2.1) is the problem of finding the tightest lower bound on the objective function that can be deduced from the constraints. It can be stated

$$\max_{v, P} \left\{ v \mid C(x) \overset{P}{\vdash} \left(f(x)v \right), \ v \in \mathbb{R}, \ P \in \mathcal{P} \right\} \tag{2.2}$$

The notation $C(x) \overset{P}{\vdash} (f(x) \geq v)$ indicates that proof P deduces $f(x) \geq v$ from $C(x)$, given that $x \in \mathcal{D}_x$. The domain of variable P is a family $\mathcal{P}$ of proofs, and a solution of the dual is a proof of the tightest lower bound v. The inference dual is always defined with respect to an inference method reflected by the proofs in $\mathcal{P}$.

We assume without any practical loss of generality that the primal problem (2.1) is bounded, meaning that there is a lower bound on $f(x)$ over all feasible x. We also assume that an optimal solution exists when the feasible set is nonempty, which is assured when the feasible set of a bounded problem is closed. When the feasible set is empty, we adopt the convention that the optimal value is ∞. Given this, we can observe that a *weak duality* property holds for inference duality: the inference dual always provides lower bound on the optimal value of the primal problem.

Lemma 2.1 *If* (v, P) *is a feasible solution of the inference dual* (2.2), *then* v *is a lower bound on the optimal value of the primal problem* (2.1).

Proof If (2.1) is feasible, it has an optimal solution x^*, and we know from the dual feasibility of (v, P) that proof P deduces $f(x^*) \geq v$. If (2.1) is infeasible, its optimal value of ∞ is bounded below by any v. $\square$

The inference dual satisfies *strong duality* for a given inference method when the optimal values of the primal and dual are necessarily equal. A solution of a strong inference dual always proves the optimality of a solution of the primal problem, or infeasibility if the problem is infeasible. However, a solution of the inference dual can often prove optimality or infeasiblity even when the dual is not strong.

Almost all optimization duals are inference duals. Consider, for example, the linear programming (LP) problem

$$\min_{x} \left\{ cx \mid Ax \geq b, \; x \geq 0 \right\} \tag{2.3}$$

The classical LP dual of (2.3) is equivalent to an inference dual that is defined with respect to nonnegative linear combinations. That is, we say that $cx \geq v$ can be inferred from $Ax \geq b$ and $x \geq 0$ when the linear combination $uAx \geq ub$ dominates $cx \geq v$ for some $u \geq 0$, where domination means that $uA \leq c$ and $ub \geq v$. Given this inference method, the inference dual (2.2) becomes

$$\max_{v,u} \left\{ v \mid uA \leq c, \; ub \geq v \text{ for some } u \geq 0 \right\}$$

which is identical to the classical LP dual $\max_u \{ub \mid uA \leq c, \; u \geq 0\}$. Similar arguments show that the Lagrangian and surrogate duals are inference duals [122, 123]. The subadditive and superadditive duals of integer programming [197] are (strong) inference duals with respect to an inference method based on Chvátal-Gomery cuts [56, 123].

For purposes of Benders decomposition, any method that solves (2.1) to optimality can be regarded as solving an inference dual. This is because the solution method can deliver a provably optimal solution x^* only by somehow establishing a bound on $f(x)$ that is equal to $f(x^*)$. The procedure used to establish this bound can be regarded as a proof that solves the inference dual, because no bound larger than $f(x^*)$ can be inferred from the constraints. If (2.4) is found to be infeasible, the procedure that establishes infeasibility can likewise be regarded as a solution of the inference dual. For example, if a problem is solved by branching, the branching tree that proves optimality or infeasibility can be regarded as a solution of the inference dual.

2.3   The LBBD Algorithm

Logic-based Benders decomposition is applied to a problem of the form

$$\min_{x,y,w} \left\{ f(x, y) \mid C(x, y),\ C'(x, w),\ x \in \mathcal{D}_x,\ y \in \mathcal{D}_y,\ w \in \mathcal{D}_w \right\} \qquad (2.4)$$

where variables $x = (x_1, \ldots, x_n)$ have been selected as master problem variables and $y = (y_1, \ldots, y_m)$ as subproblem variables. $C(x, y)$ is a set of constraints that contain variables in y and possibly in x, while the constraints in $C'(x, w)$ contain variables in x and/or w. As before, $\mathcal{D}_x$, $\mathcal{D}_y$ and $\mathcal{D}_w$ are the variable domains. The auxiliary variables w do not appear in a traditional Benders scheme but may be included in the master problem, even though they have no role in defining the subproblem. They can be useful for modeling purposes, and they allow some of the subproblem constraints to be included in the master problem, which can improve algorithmic performance (Sect. 4.6.1).

Fixing x to a given value $\bar{x}$ yields the *subproblem*

$$\min_{y} \left\{ f(\bar{x}, y) \mid C(\bar{x}, y),\ y \in \mathcal{D}_y \right\}$$

which we denote SP($\bar{x}$). The inference dual of SP($\bar{x}$) is

$$\max_{v,P} \left\{ v \mid C(\bar{x}, y) \overset{P}{\vdash} \left(f(\bar{x}, y) \geq v \right),\ v \in \mathbb{R},\ P \in \mathcal{P} \right\}$$

Let $v^*(\bar{x})$ be the optimal value of SP($\bar{x}$), and suppose that proof P^* solves the inference dual by deducing the lower bound $v = v^*(\bar{x})$ on $f(\bar{x}, y)$ for any feasible solution y of SP($\bar{x}$). The essence of LBBD is that *this same proof* may deduce, given *any* x, a useful lower bound on $f(x, y)$ for any feasible solution y of SP(x), and therefore for any feasible solution (x, y, w) of the original problem (2.4). The bound is captured in a *Benders cut* of the form $z \geq B_{\bar{x}}(x)$, where z represents the value of $f(x, y)$ in the master problem.

If $z \geq B_{\bar{x}}(x)$ is to qualify as a valid Benders cut, the bound $B_{\bar{x}}(x)$ must satisfy two properties:
(B1) $B_{\bar{x}}(x) \leq f(x, y)$ for any (x, y, w) feasible in (2.4).
(B2) $B_{\bar{x}}(\bar{x}) = v^*(\bar{x})$.

Property (B1) says that $B_{\bar{x}}(x)$ is, in fact, a valid lower bound on $f(x, y)$. Due to property (B2), the Benders cut tells the master problem that if it sends $\bar{x}$ to the subproblem again, the same optimal value $v^*(\bar{x})$ will result. This will enable us to prove finite convergence.

The Benders cut is added to the *master problem*, which contains all previously generated Benders cuts as constraints. The master problem in iteration k is

$$z_k = \min_{z,x,w} \left\{ z \mid C'(x, w);\ z \geq B_{x^\ell}(x),\ \ell = 1, \ldots, k-1;\ x \in \mathcal{D}_x,\ w \in \mathcal{D}_w \right\} \qquad (2.5)$$

where $(x^1, w^1), \ldots, (x^{k-1}, w^{k-1})$ are the solutions of the master problem in previous iterations. It is common in practice to add multiple Benders cuts in each iteration, but this has no effect on the basic theory presented here.

The next step is to solve the master problem and use its optimal solution (x^k, w^k) to define the next subproblem $SP(x^k)$. At this point, the process repeats, and it continues until the master problem has the same value as the best optimal subproblem value obtained so far; that is, until $z_k = \min\{v^*(x^\ell) \mid \ell = 1, \ldots, k-1\}$. If the master problem becomes infeasible at any point, the procedure terminates with the determination that the original problem (2.4) is infeasible. Otherwise, the procedure yields an optimal solution $(x, y, w) = (x^\ell, y^\ell, w^\ell)$ of (2.4), where ℓ is an iteration in which the best subproblem value was obtained, and y^ℓ is the optimal solution of $SP(x^\ell)$.

A formal statement of the LBBD procedure appears as Algorithm 1. Here, v^* is the best subproblem value obtained so far (∞ if no feasible subproblems have been encountered). Since z is unconstrained in the initial master problem ($k = 1$), its optimal value is $-\infty$, and any feasible (x, w) can be selected as the solution. Alternatively, we can use a "warm start" by including a few Benders cuts in the initial master problem, obtained by solving $SP(x)$ in advance for heuristically chosen values of x.

$k \leftarrow 0, v^* \leftarrow \infty$;
repeat
  $\quad k \leftarrow k + 1$;
  $\quad$ solve the master problem (2.5) and let z_k be its optimal value;
  $\quad$ **if** $z_k < \infty$ **then**
  $\quad\quad$ let (x^k, w^k) be an optimal solution of (2.5);
  $\quad\quad$ solve $SP(x^k)$, let $v^*(x^k)$ be its optimal value;
  $\quad\quad$ **if** $v^*(x^k) < v^*$ **then** $v^* \leftarrow v^*(x^k)$, $x^* \leftarrow x^k$, $w^* \leftarrow w^k$;
  $\quad\quad$ **if** $z_k < v^*$ **then** generate Benders cut $z \geq B_{x^k}(x)$ such that $B_{x^k}(x^k) = v^*(x^k)$;
  $\quad$ **end**
until $z_k = v^*$;
if $z_k < \infty$ **then** (x^*, y^*, w^*) solves (2.4), where y^* is an optimal solution of $SP(x^*)$;
else (2.4) is infeasible;

Algorithm 1: LBBD procedure when the subproblem is an optimization problem.

At any point in the algorithm, the master problem and the best subproblem value so far provide lower and upper bounds, respectively, on the optimal value of the original problem.

Lemma 2.2 *In any iteration k of Algorithm 1, we have $z_k \leq z^* \leq v^*$, where z^* is the optimal value of the original problem* (2.4).

Proof We clearly have $z^* \leq v^*$, because every subproblem solved is a restriction of (2.4). To show $z_k \leq z^*$, we recall from property (B1) that $B_{x^k}(x)$ is a valid lower bound on $f(x, y)$

for any (x, y, w) feasible in (2.4). This means that the master problem is a relaxation of (2.4), and its optimal value z_k can therefore be no greater than the optimal value z^* of (2.4). □

As the LBBD algorithm progresses, the optimal values of the master problem are monotone nondecreasing, while the optimal values of the subproblem can move up and down. This is illustrated in the example of Sect. 1.4. If the procedure is terminated early in iteration k, before achieving optimality, it nonetheless provides a lower bound x_k and upper bound v^* on the optimal value, where the upper bound is finite if at least one feasible subproblem has been encountered so far.

When a subproblem SP(x^k) is infeasible, its optimal value $v^*(x^k)$ is infinite, and the bound $B_{x^k}(x)$ is infinite for $x = x^k$. In such cases, practical algorithms replace the cut $z \geq B_{x^k}(x)$ with a *feasibility cut* $F_{x^k}(x)$ that is violated by $x = x^k$ and possibly by other values of x for which SP(x) is infeasible. In the example of Sect. 1.4, the subproblem was infeasible in iteration 1, and a feasiblilty cut $x_{11} + x_{12} \leq 1$ was generated.

The Benders procedure simplifies somewhat when the subproblem is purely a feasibility problem with no objective function, since only feasibility cuts are needed. The objective function of the original problem now depends only on the master problem variables:

$$\min_{x, y, w} \left\{ f(x) \mid C(x, y), \ C'(x, w), \ x \in \mathcal{D}_x, \ y \in \mathcal{D}_y, \ w \in \mathcal{D}_w \right\} \tag{2.6}$$

The subproblem SP(x^k) has the feasible set

$$\left\{ y \mid C(x^k, y), \ y \in \mathcal{D}_y \right\}$$

If the subprolem is infeasible, a feasibility cut $F_{x^k}(x)$ is generated. No cut is generated if the subproblem is feasible, since the procedure terminates with a solution in this case. The master problem in iteration k is

$$\min_{x, w} \left\{ f(x) \mid C'(x, w); \ F_{x^\ell}(x), \ \ell = 1, \ldots, k - 1; \ x \in \mathcal{D}_x, \ w \in \mathcal{D}_w \right\} \tag{2.7}$$

A formal statement of the procedure appears as Algorithm 2. This version of the Benders algorithm yields no feasible solution and no meaningful upper bound until it terminates, but it still provides a valid lower bound z_k on the optimal value in any iteration k.

The simplest sufficient condition for finite convergence of LBBD is that the master problem variables x_i have finite domains [126]. This is normally the case in practice, since the subproblem SP(x) is almost always defined by bounded discrete variables x_i.

Theorem 2.1 *If the domains of the master problem variables x_i are finite, Algorithm 1 and Algorithm 2 terminate after a finite number of steps with an optimal solution or by proving infeasibility.*

$k \leftarrow 0$;
repeat
  $\quad k \leftarrow k + 1$;
  $\quad$ solve the master problem (2.7);
  $\quad$ **if** (2.7) *is feasible* **then**
    $\quad\quad$ let x^k be an optimal solution of (2.5) and solve the subproblem SP(x^k);
    $\quad\quad$ **if** *SP(x^k) is infeasible* **then**
      $\quad\quad\quad$ generate a Benders cut $F_{x^k}(x)$ that is violated by $x = x^k$
    $\quad\quad$ **end**
  $\quad$ **end**
until *SP(x^k) is feasible or (2.7) is infeasible*;
if (2.7) *is feasible* **then**
  $\quad$ (x^k, y^k, w^k) solves (2.6), where y^k is an optimal solution of SP(x^k)
end
else (2.6) is infeasible;

Algorithm 2: LBBD procedure when the subproblem is a feasibility problem.

Proof It suffices to prove the theorem for Algorithm 1, because Algorithm 2 is a special case of Algorithm 1 in which the subproblem objective $f(x^k, y)$ is either 0 or ∞. We first observe that if the algorithm terminates at step k, it correctly concludes that (2.4) is infeasible, or else exhibits an optimal solution. If $z_k = \infty$, the original problem (2.4) is infeasible because $z_k \leq z^*$ by Lemma 2.2, and the algorithm correctly indicates infeasiblity. If $z_k < \infty$, then Lemma 2.2 states that $z_k \leq z^* \leq v^*$. But the condition for termination is that $z_k = v^*(x^\ell)$ for some $\ell < k$, which means $z_k = v^* = v^*(x^\ell)$ and therefore $z^* = v^*(x^\ell)$. Thus (x^ℓ, y^ℓ, w^ℓ) is an optimal solution of (2.4), as the algorithm indicates, where y^ℓ is an optimal solution of SP(x^ℓ).

To show finite convergence, we note that whenever a Benders cut $z \geq B_{x^k}(x)$ is added to the master problem, the solution x^k can again be optimal in the master problem only if the optimal value of the master problem is at least $v^*(x^k)$, due to property (B2). Thus, if x^k were found to be optimal in the master problem of a later iteration k', the algorithm would terminate, because we would have $z_{k'} = v^*(x^k) = v^*$. This implies that no optimal solution x^k of the master problem can repeat before the final iteration. Since there are finitely many possible values of x, the algorithm must terminate after finitely many iterations. $\quad\square$

2.4   Classical Benders Decomposition

Classical Benders decomposition is applied to a problem of the form

$$\min_{x,y}\{f(x) + cy \mid g(x) + Ay \geq b, \; y \geq 0, \; x \in \mathcal{D}_x\} \tag{2.8}$$

For a given $\bar{x}$, the subproblem SP($\bar{x}$) is the linear programming (LP) problem

$$\min_{y} \left\{ f(\bar{x}) + cy \mid Ay \geq b - g(\bar{x}), \ y \geq 0 \right\} \tag{2.9}$$

Classical Benders cuts are derived from the LP dual of the subproblem, which as we noted in Sect. 2.2, is an inference dual based on nonnegative linear combination as the inference method. To write the relevant dual, we remove the constant $f(\bar{x})$ from the objective function of (2.9) to obtain the LP problem

$$\min_{y} \left\{ cy \mid Ay \geq b - g(\bar{x}), \ y \geq 0 \right\} \tag{2.10}$$

whose optimal value is $\text{SP}(\bar{x}) - f(\bar{x})$. The LP dual of (2.10) is

$$\max_{u} \left\{ u(b - g(\bar{x})) \mid uA \leq c, \ u \geq 0 \right\} \tag{2.11}$$

We first suppose that (2.9) and therefore (2.10) is feasible. This means there is a finite optimal solution $\bar{u}$ of the dual (2.11). LP duality theory tells us that the primal and dual have the same optimal value (strong duality), so that

$$\bar{u}(b - g(\bar{x})) = v^{*}(\bar{x}) - f(\bar{x})$$

This allows us to write

$$f(\bar{x}) + \bar{u}(b - g(\bar{x})) = v^{*}(\bar{x})$$

Thus if we let $B_{\bar{x}}(x) = f(x) + \bar{u}(b - g(x))$ in a Benders cut $z \geq B_{\bar{x}}(x)$, the bound $B_{\bar{x}}(x)$ satisfies property (B2). The key observation for classical Benders decomposition is that $\bar{u}$ remains feasible in dual of $\text{SP}(x)$ for *any* x, since x occurs only in the objective function of the dual. Thus by weak duality,

$$\bar{u}(b - g(x)) \leq cy$$

for any x and any feasible y, and we can write

$$f(x) + \bar{u}(b - g(x)) \leq f(x) + cy \tag{2.12}$$

for any (x, y) feasible in (2.8). In other words, the same proof that establishes optimality of the subproblem $\text{SP}(\bar{x})$, namely the linear combination based on multipliers $\bar{u}$, proves a valid bound for other values of x. Due to (2.12), $B_{\bar{x}}(x)$ satisfies property (B1), and we conclude that the classical Benders cut

$$z \geq f(x) + \bar{u}(b - g(x))$$

is indeed a valid Benders cut when the subproblem is feasible.

We now suppose the subproblem (2.9) is infeasible. Classical theory (the Farkas Lemma) tells us that (2.9) is infeasible if and only if there are multipliers $\bar{u} \geq 0$ for which $\bar{u}A \leq 0$ and $\bar{u}(b - g(\bar{x})) > 0$. These same multipliers prove that any x for which $\bar{u}(b - g(x)) > 0$

gives rise to an infeasible subproblem SP(x) and is therefore infeasible in the original problem (2.8). We conclude that the classical cut for an infeasible subproblem, namely

$$\bar{u}\big(b - g(x)\big) \leq 0$$

is a valid Benders cut.

2.5   Alternative Perspectives on LBBD

One interpretation of LBBD is that it gradually computes the projection of the *epigraph* of the optimization problem (2.4) onto the master problem variables. The epigraph is

$$\big\{(z, x, y, w) \mid z \geq f(x, y), \ C(x, y), \ C'(x, w), \ w \in \mathcal{D}_x, \ y \in \mathcal{D}_y, \ w \in \mathcal{D}_w\big\}$$

The projection onto the master problem variables z, x, w is

$$\big\{(z, x, w) \mid z \geq f(x, y), \ C(x, y), \ C'(x, w), \ w \in \mathcal{D}_x, \ y \in \mathcal{D}_y, \ w \in \mathcal{D}_w\big\}$$

The Benders cuts are valid cuts for the projection. Ideally, these cuts will accurately delineate the region of the projection near the optimum before many cuts have been generated.

Still another perspective is that LBBD is an exact version of a large neighborhood search heuristic. Fixing the master problem variables defines a large neighborhood that is searched by the subproblem solver. Re-solving the master problem redefines the neighborhood. In a typical large neighborhood search, the search involves different variables in each iteration, whereas this is not the case in classical Benders decomposition. However, it will be seen in Sect. 4.8 that the choice of master problem variables in LBBD can vary from one iteration to the next, which results in a procedure that is closer to large neighborhood search.

Dynamic partitioning of master and subproblem variables can also be found in satisfiability (SAT) solvers, which can be viewed as implementing a special case of LBBD. The conflict clauses that are key to the remarkable success of these solvers are logic-based Benders cuts obtained from a conflict graph that represents a solution of the inference dual (as explained in Sects. 3.8 and 4.8). In fact, the Benders algorithm for circuit verification in [127], mentioned earlier as apparently the first occurrence of LBBD, used conflict clauses about a year before they were introduced into SAT solvers [179].

A similar interpretation can be given to solution methods for SAT modulo theories. The theory solver in these methods can be viewed as solving a LBBD subproblem and generating explanations of infeasiblity that are, in effect, Benders cuts. This is observed in a recent study of potential applications of SAT modulo theories in process engineering [184].

Finally, in an era of intense interest in machine learning as an optimization tool, one must acknowledge that Benders decomposition is none other than a very early and very successful learning method. Benders cuts record what was learned from the solution of previous subproblems, and the master problem uses this information to guide its selection of the next subproblem to solve. From this perspective, it is entirely appropriate that the SAT literature refers to the generation of conflict clauses (i.e., Benders cuts) as conflict-directed clause learning.

2.6   Subproblems That Decouple

One of the primary motivations for LBBD is to take advantage of subproblems that decouple into smaller component problems. Fortunately, convergence is still guaranteed when Benders cuts are generated for the individual component problems rather than for the subproblem as a whole.

For purposes of LBBD, an optimization problem (2.4) decouples when it can be written

$$\min_{x,y,w} \left\{ f(x,y) \;\middle|\; \begin{array}{l} C_i(x,y^i), \; y^i \in \mathcal{D}_y, \; i = 1, \ldots, m \\ C'(x,w), \; x \in \mathcal{D}_x, \; w \in \mathcal{D}_w \end{array} \right\} \tag{2.13}$$

where $y = (y^1, \ldots, y^m)$ and

$$f(x,y) = \phi\big(f_1(x, y^1), \ldots, f_m(x, y^m)\big)$$

where ϕ is a monotone nondecreasing function. For a given $\bar{x}$, the subproblem decouples into subproblems $SP_i(x)$ for $i = 1, \ldots, m$, where each subproblem i has the form

$$\min_y \left\{ f_i(\bar{x}, y^i) \;\middle|\; C_i(\bar{x}, y^i), \; y^i \in \mathcal{D}_y \right\}$$

The master problem in iteration k is

$$\min_{z,z,x,w,} \left\{ z \;\middle|\; \begin{array}{l} z \geq h(z_1, \ldots, z_m) \\ z_i \geq B^i_{x^\ell}(x), \; \ell = 1, \ldots, k-1, \; i = 1, \ldots, m \\ C'(x,w), \; x \in \mathcal{D}_x, \; w \in \mathcal{D}_w \end{array} \right\} \tag{2.14}$$

where $z_i \geq B^i_{x^\ell}(x)$ is the Benders cut generated for component i in iteration ℓ. A formal approch to checking for separability is presented in [36, 37].

In most applications where the subproblem decouples, $f(x,y)$ is additively or multiplicatively separable, so that

$$f(x,y) = \phi\big(f_1(x, y^1), \ldots, f_m(x, y^m)\big) = \begin{cases} \sum_i f_i(x, y^i) \\ \text{or } \prod_i f_i(x, y^i) \end{cases}$$

In the assignment and scheduling example of Sect. 1.4, $f(x,y)$ is not separable, but we still have decoupling with ϕ as a maximum function. Here, $f_i(x, y^i)$ is the makespan in shop i, and $f(x,y)$ the overall makespan, so that

$$f(x, y) = \phi\big(f_1(x, y^1), \ldots, f_m(x, y^m)\big) = \max_i \big\{ f_i(x, y^i) \big\}$$

All three of these functions ϕ are nondecreasing.

To ensure convergence, we need only design cuts that satisfy (B1) and (B2) for individual subroblem components. That is, it suffices that

(B1)$_i$ $B_{\bar{x}}^i(x) \le f_i(x, y^i)$ for any (x, y, w) feasible in (2.13).

(B2)$_i$ $B_{\bar{x}}^i(\bar{x}) = v_i^*(\bar{x})$.

for each i, where $v_i^*(x)$ is the optimal value of SP$_i(x)$.

Corollary 2.1 *Suppose Algorithm 1 is applied to (2.13) by generating a master problem of the form (2.14) rather than (2.5). Then if the master problem variables have finite domains, the algorithm terminates after a finite number of steps with an optimal solution or by proving infeasibility.*

Proof Due to Theorem 2.1, it suffices to show: (a) the constraints

$$z \ge \phi(z_1, \ldots, z_m), \text{ and } z_i \ge B_{\bar{x}}^i(x) \text{ for } i = 1, \ldots, m \tag{2.15}$$

are equivalent to the cut $z \ge B_{\bar{x}}(x)$, where

$$B_{\bar{x}}(x) = \phi(B_{\bar{x}}^1(x), \ldots, B_{\bar{x}}^m(x)) \tag{2.16}$$

and (b) $B_{\bar{x}}(x)$ satisfies (B1) and (B2). To show (a), suppose first that x, z, and $(z_1, \ldots, z_m)$ satisfy (2.15). Due to $z \ge \phi(z_1, \ldots, z_m)$ and (2.16), the cut $z \ge B_{\bar{x}}(x)$ follows from

$$\phi(z_1, \ldots, z_m) \ge \phi(B_{\bar{x}}^1(x), \ldots, B_{\bar{x}}^m(x))$$

But this holds because ϕ is nondecreasing and $z_i \ge B_{\bar{x}}^i(x)$ for all i. Now we suppose that x and z satisfy $z \ge B_{\bar{x}}(x)$. It suffices to show that (2.15) holds for some tuple $(z_1, \ldots, z_m)$. But for this we need only let $z_i = B_{\bar{x}}^i(x)$ for each i. This proves (a). To prove (b), we first show that $B_{\bar{x}}(x)$ satisfies (B1), which is equivalent to showing

$$\phi\big(B_{\bar{x}}^1(x), \ldots, B_{\bar{x}}^m(x)\big) \le \phi\big(f_1(x, y^1), \ldots, f_m(x, y^m)\big)$$

for all feasible (x, y). But this holds because ϕ is nondecreasing and $B_{\bar{x}}^i(x) \le f_i(x, y^i)$ for all i due to (B1)$_i$. Finally, we show that $B_{\bar{x}}(x)$ satisfies (B2) by showing

$$\phi\big(B_{\bar{x}}^1(x), \ldots, B_{\bar{x}}^m(x)\big) = v^*(x) \tag{2.17}$$

But since ϕ is nondecreasing, we know $v^*(x) = h\big(v_1^*(x), \ldots, v_m(x)\big)$, and so (2.17) follows from the premise that (B2)$_i$ is satisfied for all i. $\qquad\square$

In the assignment and scheduling example, the cut $M \geq 4x_{24}$ we generated for shop 2 in the first iteration is not a valid Benders cut $z \geq B_{\bar{x}}(x)$ for the entire subproblem, because it violates (B2). Here (B2) requires $4\bar{x}_{24} = v^*(\bar{x})$, which is violated because $\bar{x}_{24} = 1$ and $v^*(\bar{x}) = \infty$; the minimum makespan $v^*(\bar{x})$ for the subproblem as a whole is infinite because there is no feasible schedule in shop 1. However, if we view $M \geq 4x_{24}$ as a cut $z_2 \geq B_{\bar{x}}^2(x)$ for shop 2 alone, (B2)$_2$ is satisfied because the minimum makespan in shop 2 is $v_2^*(\bar{x}) = 4$. This is enough to ensure convergence.

2.7   Practical Guidelines

If a problem is suitable for decomposition, a properly-designed LBBD algorithm can bring dramatic reductions in computation time, enabling solution of instances that were previously intractable. The following guidelines may be of assistance in crafting an effective design.

- A problem is decomposed for purposes of LBBD by partitioning the variable set into "complicating" variables and the remaining variables. The master problem contains the complicating variables, leaving the rest for the subproblem. As the name suggests, complicating variables should be chosen so that, when fixed, the resulting subproblem is much less complicated than the original problem.
- One way to obtain a simpler subproblem is to select complicating variables that, when fixed, allow the subproblem to decouple into smaller problems that can be solved separately. However, there are many successful applications of LBBD in which the subproblem does not decouple.
- One advantage of LBBD is that the master problem and subproblem can be solved by different methods, and the decomposition can be designed with this in mind. For example, one might place constraints suitable for mixed integer programming in the master problem and constraints suitable for constraint programming or a specialized algorithm in the subproblem.
- Computational experience suggests that, for best performance, the master problem and subproblem should require roughly equal solution time, on the average, while recognizing that the master solution complexity may increase as Benders cuts are added.
- A rule of thumb is that if more than 100 or so Benders iterations are required, the LBBD approach has bogged down and should be redesigned. There is no need for redesign, of course, if a large number of iterations are proceeding rapidly to a solution.
- When the master problem is hard to solve, branch and check (Sect. 4.2) can be an attractive alternative, provided the master problem is solved by a branching method. This variant of LBBD solves the master problem only once and therefore allows the inclusion of harder constraints in the master problem.
- The design of Benders cuts can be crucial to success. Various options for formulating cuts are described in Chap. 3. Most applications to date combine strengthened nogood

cuts with some type of analytical cuts, although a number of applications succeed with nogood cuts alone, even unstrengthened ones.

- The methods used in the literature to strengthen nogood cuts can frequently be improved using the algorithms presented in Sect. 3.2.4.
- The greatest potential for cut strengthening lies in the formulation of analytical cuts that take advantage of special structure in the subproblem. Ideas for effective analytical cuts can be found in Sects. 3.3–3.6, as well as in the many articles summarized in Chap. 5.
- Classical Benders cuts can, of course, be used if the subproblem is a linear programming problem. However, even in this case, it can be helpful to apply branch and check with combinatorial Benders cuts (Sect. 4.4) if the master problem is a mixed integer programming problem.
- LBBD is particularly attractive for two-level stochastic and robust optimization, because the second stage problem can be viewed as a Benders subproblem that decouples into a smaller problem for each scenario (Sect. 4.5).
- It is important in many applications to include a subproblem relaxation in the master problem, particularly if the Benders cuts do not convey much information about the subproblem. Strategies for accomplishing this are discussed in Sect. 4.6.
- When both the master problem and subproblem are too hard so solve, multilevel LBBD has occasionally been found useful (Sect. 4.7).
- The partition of the variable set can vary from one iteration to the next. This option is discussed in Sect. 4.8.
- Another technique for strengthening the master problem, long used in classical Benders decomposition, is a "warm start." This involves generating a set promising solutions heuristically and submitting them to the subproblem to generate cuts that are included the master problem at the outset.
- When solution of the subproblem is time-consuming, one popular option is to solve an easier relaxation of the subproblem first. If the relaxation is infeasible, Benders cuts can be generated, whereupon the algorithm proceeds to the next iteration without solving the full subproblem. If the relaxed subproblem is feasible, the full subproblem is solved.
- Although the development of general-purpose LBBD solvers is currently at an early stage, a solver of this sort may deliver significantly better performance than a standard method, even if not up to the level that a hand-crafted decomposition and specialized cuts would provide (Sects. 3.8 and 4.8).

Logic-Based Benders Cuts

3

3.1 Introduction

There are three general categories of logic-based Benders cuts. *Nogood cuts* are derived solely from optimal values returned by the subproblem. An important subcategory consists of *monotone cuts*, which are obtained from a subproblem whose optimal value is a nondecreasing function of the master problem solution. Various algorithms are available to strengthen monotone cuts by intelligently sampling certain subproblem solutions. We describe these algorithms in the first major section of the chapter and illustrate them with examples.

Analytical cuts are derived by combining the current subproblem solution with an analysis of the subproblem structure. These cuts are often more effective than strengthened nogood cuts because they reflect a deeper understanding of problem characteristics. We first describe a class of generic analytical optimality cuts. Following this, we develop analytical cuts for scheduling problems at some length, since this is the most intensively studied application domain for LBBD. In so doing, we illustrate some of the strategies that can be used to derive cuts in other domains. We also derive analytical cuts for vehicle routing problems and packing problems, as further examples.

The third category consists of *explanation-based cuts*, which are closest to the original motivation for LBBD. They are derived directly from the inference dual solution of the subproblem, which is to say, from the subproblem's proof of optimality or infeasibility. Since few optimization solvers currently provide this kind of explanation, it is usually necessary to infer proof characteristics by sampling subproblem solutions or analyzing subproblem structure. This indirect strategy has nonetheless achieved much success, as nearly all applications to date rely on nogood cuts and analytical cuts. A few solvers are beginning to supply explanations, particularly propositional satisfiability (SAT) and constraint programming solvers, and we show in the final section how these explanations can form the basis for logic-based cuts.

J. Hooker, *Logic-Based Benders Decomposition*, Synthesis Lectures on Operations Research and Applications, https://doi.org/10.1007/978-3-031-45039-6_3

Table 3.1 Benders cut nomenclature used in this book

Feasibility cut	A cut that is derived from an infeasible subproblem and excludes one or more infeasible assignments to the master problem variables
Optimality cut	A cut that is derived from a feasible subproblem and places a bound on the objective function value that results from one or more assignments to the master problem variables
Classical Benders cut	A Benders cut in the form of an inequality that is based on dual multipliers obtained by solving the dual of a linear programming subproblem
Simple nogood cut	A logic-based cut that excludes only the current value assignments to master problem variables, or that provides a meaningful bound only for these value assignments
Simple feasiblity (or optimality) cut	A simple nogood cut that is a feasiblity (or optimality) cut
Simple monotone cut	A nogood cut obtained from a subproblem whose optimal value is a monotone function of the master problem variables, and that therefore excludes or provides a bound for many more variable assignments than a simple nogood cut
Strengthened monotone cut	A monotone feasibility or optimality cut from which one or more master problem variables have been removed without sacrificing validity, allowing the cut to exclude or provide a bound for many more variable assignments than a simple monotone cut
Multivalent cut	A optimality nogood cut that can provide a weaker but meaningful bound on the objective function when the values of one or more master problem variables in the cut are changed from their current values in the master problem solution
Analytical cut	A strengthened feasibility or optimality cut that is based on structural analysis of the subproblem and its current optimal solution (or lack of a feasible solution), rather than on observing the effect on the subproblem solution of removing one or more variables from the cut
Explanation-based cut	A feasibility or optimality cut that is derived directly from an inference dual solution of the subproblem (provided by the suproblem solver), rather than indirectly as in the case of a strengthened nogood cut or an analytical cut
Enumerative cut	A feasibility cut, derived from a feasible rather than an infeasible subproblem, that excludes assignments to master problem variables that result in a subproblem value no better than the current value
Combinatorial Benders cut	A nogood feasibility cut or enumerative cut, derived from a linear programming subproblem, that is used in the context of a branch-and-cut method applied to a mixed integer/linear programming problem

While logic-based cuts in the literature can be highly problem-specific, most of them follow patterns similar to those of cut families presented in this chapter. In any event, Chap. 5 provides brief descriptions of a wide variety of cuts that have been used in practice and directs readers to the relevant articles. Table 3.1 provides a glossary of cut nomenclature used in this and subsequent chapters.

3.2   Nogood Cuts

Nogood cuts are based on the idea that if a given solution $\bar{x}$ of the master problem results in an infeasible subproblem, the assignment $x = \bar{x}$ is "no good," and a Benders cut can exclude it from future master problem solutions. If the subproblem is feasible with optimal value $v^*(\bar{x})$, a nogood cut imposes a lower bound of $v^*(\bar{x})$ on the objective function value when $x = \bar{x}$. This tells the master problem that sending the same value $\bar{x}$ to the subproblem again is "no good" because it will not improve the solution.

These simple nogood cuts are naturally very weak because they are activated by only one value $\bar{x}$ of x, but they can normally be strengthened. In many applications, the optimal subproblem value $v^*(x)$ is a monotone nondecreasing function of x, which already permits a significantly stronger cut. The cut can often be further strengthened by observing which variables x_i have no effect on the optimal subproblem value, and removing them from the cut. Various algorithms are available to identify such superfluous variables, some of which yield irreducible cuts (i.e., no further variables can be removed without reducing the lower bound imposed by the cut). Other algorithms ascertain how much the valid lower bound degrades when a certain number of variables are removed, and one writes a cut on this basis. Both types of algorithms require solving multiple subproblems to create a single cut, but this approach often works quite well in practice due to rapid solution of subproblems.

In nearly all applications, the master problem variables x_j that define the subproblem are binary. Typically, $x_j = 1$ when a task j is must be included in the subproblem. We therefore indicate how the various cuts are formulated in terms of 0–1 variables. In addition, the master problem is often a mixed integer/linear programming (MILP) problem, which requires linear constraints. We therefore present a linearized cut for binary variables x_j.

3.2.1   Simple Nogood Cuts

The most straightforward type of nogood cut is a *simple feasibility cut*, which excludes a value $\bar{x}$ of x that has been found to create an infeasible subproblem $SP(\bar{x})$. The cut is therefore simply $x \neq \bar{x}$. When x consists of binary variables, the linearized cut is

$$\sum_{j \in J_0(\bar{x})} x_j + \sum_{j \in J_1(\bar{x})} (1 - x_j) \geqslant 1 \tag{3.1}$$

where $J_0(x) = \{j \mid x_j = 0\}$ and $J_1(x) = \{j \mid x_j = 1\}$. If the subproblem is feasible, an analogous *simple optimality cut* imposes the lower bound $v^*(\bar{x})$ when $x = \bar{x}$ but no lower bound otherwise. For binary x, the linearized cut is

$$z \geqslant v^*(\bar{x}) - \left(v^*(\bar{x}) - \underline{v}\right) \left(\sum_{j \in J_1(\bar{x})} (1 - x_j) + \sum_{j \in J_0(\bar{x})} x_j \right) \tag{3.2}$$

Fig. 3.1 A minimum
makespan schedule

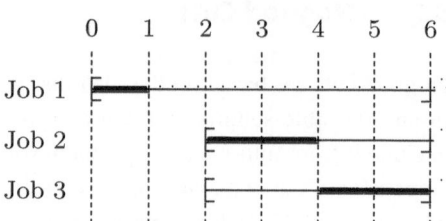

where $\underline{v}$ is a lower bound on $v^*(x)$ that is valid for all x that are feasible in the master problem. The right-hand side of (3.2) imposes the lower bound $v^*(x)$ when $x = \bar{x}$ and an innocuous lower bound of at most $\underline{v}$ when x differs from $\bar{x}$ in one or more components. It forces the master problem to try a value of x different from $\bar{x}$ in order to find a better solution. We refer to both (3.1) and (3.2) as *simple nogood cuts*.

A small example similar to that given in Chap. 1 will illustrate several cuts described in this chapter. We wish to use LBBD to assign four jobs to shops and schedule them in those shops, so as to minimize makespan (the finish time of the last job to finish). We suppose that the current solution of the master problem assigns jobs 1, 2 and 3 to a certain shop, so that $\bar{x} = (\bar{x}_1, \ldots, \bar{x}_4) = (1, 1, 1, 0)$. An optimal schedule on this shop is depicted in Fig. 3.1, where the brackets again indicate time windows and the heavy horizontal lines indicate processing times, and the minumum makespan is $v^*(\bar{x}) = 6$. We suppose further that the master problem requires that at least one of these three jobs be assigned to this shop, which means that the makespan is bounded below by $\underline{v} = 1$. Since this subproblem is feasible, we can write a simple feasibility nogood cut (3.2):

$$z \geqslant 6 - 5\big((1 - x_1) + (1 - x_2) + (1 - x_3) + x_4\big) \qquad (3.3)$$

The cut is useful only when $x = (1, 1, 1, 0)$.

3.2.2   Cuts with General Integer Variables

Before continuing with a survey of types of nogood cuts, we pause to indicate how one might write linear formulations of cuts that contain general integer variables x_j. It can be done in at least two ways by introducing 0–1 variables. One method is to let 0–1 variable δ_{jt} take the value 1 when $x_j = t$, where t ranges over the possible integer values of x_j. This is achieved by adding the constraints

$$x_j = \sum_t t \delta_{jt}, \quad \text{all } j$$

to the master problem at the beginning of the LBBD algorithm. Then since a nogood cut $x \neq \bar{x}$ can be written as a disjunction

$$\bigvee_j (x_j \neq \bar{x}_j)$$

any such cut can be given the 0–1 formulation

$$\sum_j \left(1 - \delta_{j\bar{x}_j}\right) \geqslant 1$$

This may be practical if the range of x_j is small.

If the desired cuts have the form

$$\bigvee_{j \in J} (x_j \geqslant \bar{x}_j)$$

as frequently occurs, it enough to introduce an indicator variable $\delta_{j\bar{x}_j}$ for each $j \in J$, such that $\delta_{j\bar{x}_j} = 1$ when $x_j < \bar{x}_j$. This is done by adding the big-M constraints

$$x_j \geqslant \bar{x}_j - M\delta_{j\bar{x}_j}, \quad j \in J$$

to the master problem (where M is a large number), whereupon the cut can be written

$$\sum_{j \in J} (1 - \delta_{j\bar{x}_j}) \geqslant 1$$

This method may be preferred when x_j has a large domain, since we need only $|J|$ 0–1 variables for a given cut. However, it may be necessary to introduce new 0–1 variables as each cut is generated. Specifically, $\delta_{j\bar{x}_j}$ must be introduced for each $j \in J$, unless the current value of $\bar{x}_j$ is the same as in some previous cut. In practice, it may be preferable to solve the master problem by some method other than MILP when the cuts involve general integer variables, such as constraint programming.

3.2.3   Monotone Nogood Cuts

As noted above, simple nogood cuts are extremely weak because they are activated by a single value of x. Cuts based on the same idea are stronger when $v^*(x)$ is a monotone nondecreasing function of x; that is, $x \geqslant \bar{x}$ implies $v^*(x) \geqslant v^*(\bar{x})$. In particular, monotonicity means that increasing one or more components of $\bar{x}$ preserves infeasiblity, because $v^*(x) = \infty$ if $x \geqslant \bar{x}$ and $v^*(\bar{x}) = \infty$. Thus if SP$(\bar{x})$ is infeasible, a valid feasiblity cut $F_{\bar{x}}(x)$ can rule out any x for which $x \geqslant \bar{x}$. We refer to this as a *monotone feasbility cut*. For binary x, the linearized cut is a stronger version of (3.1):

$$\sum_{j \in J_1(\bar{x})} (1 - x_j) \geqslant 1$$

A *monotone optimality cut* says that $v^*(\bar{x})$ becomes no smaller when $\bar{x}$ is replaced by any x satisfying $x \geqslant \bar{x}$. The binary version of the cut is similar to (3.2) but stronger:

$$z \geqslant v^*(\bar{x}) - \left(v^*(\bar{x}) - \underline{v}\right) \sum_{j \in J_1(\bar{x})} (1 - x_j)$$

Rather than excluding a single value $\bar{x}$ of x, a monotone cut excludes all values of x that dominate $\bar{x}$ and is therefore significantly stronger than a simple nogood cut.

In the example of Sect. 3.2.1, $v^*(x)$ is clearly monotone nondecreasing, because assigning an additional job to a shop cannot reduce the minimum makespan. The simple cut (3.3) can therefore be strengthened to

$$z \geqslant 6 - 5\big((1 - x_1) + (1 - x_2) + (1 - x_3)\big) \tag{3.4}$$

which is activated when x is $(1, 1, 1, 0)$ or $(1, 1, 1, 1)$.

3.2.4 Strengthened Monotone Cuts

While monotonicity allows stronger cuts, good performance in practice often requires still stronger cuts. In such cases, one option is to explore computationally (by repeatedly re-solving the subproblem) whether some of the variables x_j in the cut have no impact on the value of $v^*(x)$. If so, we can remove these variables and obtain a *strengthened monotone cut*, which can be either a feasiblity or an optimality cut. This maneuver can be viewed as identifying variables x_j that play no essential role in the proof of optimality of the subproblem solution.

For ease of presentation, we assume x is a binary vector. The ideas can be extended to general integer variables if desired. A simple monotone cut contains all the variables x_j for which $\bar{x}_j = 1$. A *strengthened feasibility cut* contains a proper subset of these variables and can be written

$$\sum_{j \in J_1(\hat{x})} (1 - x_j) \geqslant 1 \tag{3.5}$$

where $J_1(\hat{x}) \subset J_1(\bar{x})$, while a *strengthened optimality cut* has the form

$$z \geqslant v^*(\bar{x}) - \left(v^*(\bar{x}) - \underline{v}\right) \sum_{j \in J_1(\hat{x})} (1 - x_j) \tag{3.6}$$

The cut is *irreducible* if it becomes invalid when any further variables x_j are removed. That is, $v^*(x') < v^*(\hat{x})$ whenever $J_1(x')$ is a proper subset of $J_1(\hat{x})$. An irreducible cut does not necessarily have a minimum number of variables, because there may be other valid cuts that use still fewer variables.

We would like to strengthen monotone cuts by removing variables when possible, perhaps obtaining an irreducible cut. Three procedures for doing so are presented in [148]. One is a greedy procecure, and the other two (a deletion filter and a depth-first binary search procedure) yield irreducible cuts. In addition, there is a greedy procedure based on binary search, as well as the QuickXplain algorithm [142], which was originally developed for constraint programming and yields an irreducible cut. One version of QuickXplain uses binary search. These procedures are presented in the literature as methods to find small infeasible subsets of constraints, but we adapt them to identify strengthened optimality cuts as well as feasibility cuts.

Greedy procedure. This procedure simply tries flipping to 0 each $\bar{x}_j$ that is equal to 1. If the resulting subproblem still has optimal value $v^*(\bar{x})$ after $\bar{x}_j$ is flipped, we remove variable x_j from the optimality cut and continue. Otherwise we retain x_j in the cut and stop. The procedure is heuristic in the sense that it need not yield an irreducible cut. It is stated more precisely as Algorithm 1. This and subsequent procedures can be applied when the subproblem is a feasibility problem, if we recall that infeasibility occurs when $v^*(\bar{x}) = \infty$. In this case, the algorithm flips variables $\bar{x}_j$ to 0 until the resulting subproblem is feasible.

$\hat{x} \leftarrow \bar{x}$;
repeat
  select $j \in J_1(\hat{x})$;
  $\hat{x}_j \leftarrow 0$;
until $v^*(\hat{x}) < v^*(\bar{x})$ *or* $J_1(\hat{x}) = \emptyset$;
if $v^*(\hat{x}) < v^*(\bar{x})$ **then** $\hat{x}_j \leftarrow 1$;
generate cut (3.5)

Algorithm 1: Greedy procedure for strengthening a monotone optimality cut

In the example (3.4) of the previous section, the greedy procedure first flips $\bar{x}_1$ to 0. Since the optimal makepan is still 6, it removes x_1 from the cut (3.4) and tries flipping $\bar{x}_2$ to 0. Since the optimal makespan is reduced to 4, the procedure stops with the strengthened cut

$$z \geqslant 6 - 5\big((1 - x_2) + (1 - x_3)\big) \tag{3.7}$$

An example that is more adequate for the methods of this section goes as follows. Suppose $\bar{x} = (\bar{x}_1, \ldots, \bar{x}_6) = (1, 1, 1, 1, 1, 1)$, and the subproblem SP($x$) is a feasibility problem with $v^*(x)$ equal to 0 when SP(x) is feasible and equal to ∞ when it is infeasible. Suppose further that $v^*(x) = \infty$ for any x in which $x_2 = x_4 = x_5 = 1$, and $v^*(x) = 0$ for other values of x. Then there is a single irreducible monotone cut:

$$(1 - x_2) + (1 - x_4) + (1 - x_5) \geqslant 1 \tag{3.8}$$

Table 3.2 Examples of three monotone cut strengthening algorithms: (a) greedy, (b) deletion filter, and (c) heuristic binary search

(a)

j	$\hat{x}$	$v^*(\hat{x})$
	$(1,1,1,1,1,1)$	∞
1	$(0,1,1,1,1,1)$	∞
2	$(0,0,1,1,1,1)$	0
	$(0,1,1,1,1,1)$	

(b)

j	$\hat{x}$	$v^*(\hat{x})$
	$(1,1,1,1,1,1)$	∞
1	$(0,1,1,1,1,1)$	∞
2	$(0,0,1,1,1,1)$	0
3	$(0,1,0,1,1,1)$	∞
4	$(0,1,0,0,1,1)$	0
5	$(0,1,0,1,0,1)$	0
6	$(0,1,0,1,1,0)$	∞

(c)

ℓ	u	m	$\hat{x}$	$v^*(\hat{x})$
			$(1,1,1,1,1,1)$	∞
0	6	3	$(1,1,1,0,0,0)$	0
4	6	5	$(1,1,1,1,1,0)$	∞
4	5	4	$(1,1,1,1,0,0)$	0
5	5		$(1,1,1,1,1,0)$	

An implementation of the greedy procedure on this example appears in Table 3.2(a). It yields the strengthened cut

$$(1 - x_2) + (1 - x_3) + (1 - x_4) + (1 - x_5) + (1 - x_6) \geqslant 1$$

which is, however, not irreducible.

Deletion filter. This procedure (Algorithm 2) is used in [54] to find irreducible sets of linear constraints. It tries flipping to 0 *every* $\bar{x}_j$ that is equal to 1. The value of x_j is restored to 1 when the resulting subproblem has optimal value equal to $v^*(\bar{x})$. This procedure always identifies an irreducible cut but may require significantly more computation than the heuristic procedure. It solves all the subproblems indexed by $J_1(\bar{x})$, while the heuristic method stops solving them as soon as $v^*(\hat{x}) < v^*(\bar{x})$. Table 3.2(b) illustrates the deletion filter on the above example, yielding the irreducible cut (3.8).

Depth-first binary search (DFBS). This procedure accelerates the deletion filter process by dropping several variables at once from the cut, rather than one at a time, when check-

---

$\hat{x} \leftarrow \bar{x}$;
foreach $j \in J_1(\bar{x})$ **do**
$\quad | \quad \hat{x}_j \leftarrow 0$;
$\quad | \quad$ **if** $v^*(\hat{x}) < v^*(\bar{x})$ **then** $\hat{x}_j \leftarrow 1$;
end
generate cut (3.5)

Algorithm 2: Deletion filter for identifying an irreducible monotone optimality cut

ing whether the cut remains valid. If the cut is still valid, the procedure restarts with this smaller cut. If the cut is no longer valid, then it begins restoring variables to the cut until validity is achieved. This is done by binary search: first restore half the omitted variables, and if necessary half the remaining variables, and so forth until validity is restored. Whenever there is only one remaining variable, and restoring that variable achieves validity, we know that variable is part of an irreducible cut. The formal statement of the procedure in Algorithm 3 uses the notation $x(J)$ to denote a solution x in which $x_j = 1$ if $j \in J$ and $x_j = 0$ if $j \notin J$. In the algorithm, J contains the indices of variables that have so far been selected for the irreducible cut, T indexes the variables on which a binary search is being conducted, and S indexes the remaining variables aside from those indexed by J. The previous example is processed in Table 3.3, again yielding the irreducible cut (3.8).

$T \leftarrow J_1(\bar{x}); J \leftarrow \emptyset; S \leftarrow \emptyset$;
repeat
$\quad | \quad$ **if** $|T| \leqslant 1$ **then**
$\quad | \quad | \quad J \leftarrow J \cup T$;
$\quad | \quad | \quad$ **if** $v^*(x(J)) = v^*(\bar{x})$ **then** stop, generate cut (3.6) with $\hat{x} = x(J)$;
$\quad | \quad | \quad$ **else**
$\quad | \quad | \quad | \quad T \leftarrow S; S \leftarrow \emptyset$;
$\quad | \quad | \quad | \quad$ **if** $|T| \geqslant 2$ **then** split T into T_1 and T_2;
$\quad | \quad | \quad | \quad$ **else** $T_2 \leftarrow T; T_1 \leftarrow \emptyset$;
$\quad | \quad | \quad$ **end**
$\quad | \quad$ **end**
$\quad | \quad$ **else** split T into T_1 and T_2;
$\quad | \quad J \leftarrow J \cup S \cup T_1$;
$\quad | \quad$ **if** $v^*(x(J)) = v^*(\bar{x})$ **then** $T \leftarrow T_1$;
$\quad | \quad$ **else** $T \leftarrow T_2; S \leftarrow S \cup T_1$;
$\quad | \quad J \leftarrow J \setminus (S \cup T_1)$
end

Algorithm 3: DFBS procedure for identifying an irreducible monotone optimality cut

Heuristic binary search. The binary search idea of DFBS can be used in a simpler and faster heuristic algorithm, proposed by [21], that does not guarantee an irreducible cut. We suppose that $\bar{x} = (\bar{x}_1, \ldots, \bar{x}_n)$ with $\bar{x}_i = 1$ for all j. In addition, the variables $x_1, \ldots, x_n$

Table 3.3 Example of a depth-first binary search (DFS) algorithm for monotone cut strengthening

T	J	S	T_1	T_2	J	$v^*(x(J))$
$\{1, 2, 3, 4, 5, 6\}$	$\emptyset$	$\emptyset$	$\{1, 2, 3\}$	$\{4, 5, 6\}$	$\{1, 2, 3\}$	0
$\{4, 5, 6\}$	$\emptyset$	$\{1, 2, 3\}$	$\{4, 5\}$	$\{6\}$	$\{1, 2, 3, 4, 5\}$	∞
$\{4, 5\}$	$\emptyset$	$\{1, 2, 3\}$	$\{4\}$	$\{5\}$	$\{1, 2, 3, 4\}$	0
$\{5\}$	$\emptyset$	$\{1, 2, 3, 4\}$			$\{5\}$	0
$\{1, 2, 3, 4\}$	$\{5\}$	$\emptyset$	$\{1, 2\}$	$\{3, 4\}$	$\{1, 2, 5\}$	0
$\{3, 4\}$	$\{5\}$	$\{1, 2\}$	$\{3\}$	$\{4\}$	$\{1, 2, 3, 5\}$	0
$\{4\}$	$\{5\}$	$\{1, 2, 3\}$			$\{4, 5\}$	0
$\{1, 2, 3\}$	$\{4, 5\}$	$\emptyset$	$\{1, 2\}$	$\{3\}$	$\{1, 2, 4, 5\}$	∞
$\{1, 2\}$	$\{4, 5\}$	$\emptyset$	$\{1\}$	$\{2\}$	$\{1, 4, 5\}$	0
$\{2\}$	$\{4, 5\}$	$\{1\}$			$\{2, 4, 5\}$	∞

are ordered by nonincreasing relevance, meaning that variables near the front of the list are more likely to be part of an irreducible cut. The algorithm iteratively updates a lower bound ℓ and upper bound u on the number of variables in an irreducible cut, where initially $\ell = 0$ and $u = n$. A midpoint index m between ℓ and u is computed, and the subproblem is re-solved with $\bar{x}_{m+1}, \ldots, \bar{x}_n$ set to zero. If the value of the subproblem is unchanged, we observe that these variables can be dropped from the cut, update u to m, and repeat. If the value of the subproblem is reduced, we must reduce the number of variables to be dropped by updating ℓ to $m + 1$, and repeat. The procedure stops when $\ell = u$, at which point we have a strengthened cut that contains variables $x_1, \ldots, x_\ell$. A formal statement of the procedure appears as Algorithm 4. The example in Table 3.2(c) yields the cut

$$(1 - x_1) + (1 - x_2) + (1 - x_3) + (1 - x_4) + (1 - x_5) \geqslant 1$$

which is not irreducible.

$\ell \leftarrow 0; u \leftarrow n;$
while $\ell < u$ **do**
    $m \leftarrow \ell + \lfloor \frac{1}{2}(u - \ell) \rfloor;$
    $\hat{x}_j = 1$ for $j = 1, \ldots, m;$ $\hat{x}_j = 0$ for $j = m + 1, \ldots, n;$
    if $v^*(\hat{x}) = v^*(\bar{x})$ **then** $u \leftarrow m;$
    else $\ell \leftarrow m + 1;$
end
$\hat{x}_j = 1$ for $j = 1, \ldots, \ell;$ $\hat{x}_j = 0$ for $j = \ell + 1, \ldots, n;$
generate cut (3.5)

Algorithm 4: Heuristic binary search for strengthening a monotone optimality cut, where $\bar{x} = (\bar{x}_1, \ldots, \bar{x}_n)$ with $\bar{x}_j = 1$ for $j = 1, \ldots, n$

Table 3.4 Example of QuickXplain without binary search

T	U	m	k	C	$v^*(x(C))$
$\emptyset$	$\{1, 2, 3, 4, 5, 6\}$	6	0	$\emptyset$	0
			1	$\{1\}$	0
			2	$\{1, 2\}$	0
			3	$\{1, 2, 3\}$	0
			4	$\{1, 2, 3, 4\}$	0
			5	$\{1, 2, 3, 4, 5\}$	∞
$\{5\}$	$\{1, 2, 3, 4\}$	4	0	$\{5\}$	0
			1	$\{5, 1\}$	0
			2	$\{5, 1, 2\}$	0
			3	$\{5, 1, 2, 3\}$	0
			4	$\{5, 1, 2, 3, 4\}$	∞
$\{4, 5\}$	$\{1, 2, 3\}$	3	0	$\{4, 5\}$	0
			1	$\{4, 5, 1\}$	0
			2	$\{4, 5, 1, 2\}$	∞
$\{2, 4, 5\}$	$\{1\}$	1	0	$\{2, 4, 5\}$	∞

QuickXplain. This procedure, introduced in [142], is likewise based on a priority ordering of the variables $x_1, \ldots, x_n$ in the cut. It can be used with or without binary search. If binary search is not used, the algorithm proceeds as follows. Variables $x_1, x_2, \ldots$ are added one by one to the cut, each time testing whether the optimal subproblem value is unchanged. We take note of the first variable for which the subproblem value is unchanged and let set T consist of this variable. In the next iteration, we start with a cut consisting of the variable(s) in T, and again take note of the first variable that leaves the subproblem value unchanged, and add it to T. We continue until the variables in T alone leave the subproblem value unchanged, whereupon we have an irreducible cut containing these variables. The procedure appears as Algorithm 5. Table 3.4 applies QuickXplain to the above example, yielding the irreducible cut (3.8).

A binary search element can be added by splitting the variables in x other those in T into two subsets, and applying the procedure recursively to each subset. The formal procedure appears as Algorithm 6. The example displayed in Table 3.5 begins with the top level of the recursion by computing $J = \text{QuickXplain}(C, U)$ with $C = \emptyset$ and $U = \{1, 2, 3, 4, 5, 6\}$, which returns $J = \{2, 4, 5\}$ and produces the irreducible cut (3.8). The computation of QuickXplain(C, U) makes recursive calls to QuickXplain(C, U_2) and then QuickXplain(C, U_1), necessarily in that order. In the first call, $C = \{5, 1, 2\}$ and $U_2 = \{3, 4\}$, whereas $C = \{4, 5\}$ and $U_1 = \{1, 2\}$ in the second call. The table also displays the calculation of QuickXplain(C, U_2) and QuickXplain(C, U_1).

Table 3.5 Example of QuickXplain with binary search

k	C	$v^*(C)$	T	U_1	U_2
Compute QuickXplain(∅, {1, 2, 3, 4, 5, 6})					
0	∅	0			
1	{1}	0			
2	{1, 2}	0			
3	{1, 2, 3}	0			
4	{1, 2, 3, 4}	0			
5	{1, 2, 3, 4, 5}	∞	{5}	{1, 2}	{3, 4}
$T \leftarrow \{5\} \cup$ QuickXplain({5, 1, 2}, {3, 4}) = {5} ∪ {4} = {4, 5}					
$T \leftarrow \{4, 5\} \cup$ QuickXplain({4, 5}, {1, 2}) = {4, 5} ∪ {2} = {2, 4, 5}}					
Compute QuickXplain({5, 1, 2}, {3, 4})					
0	{5, 1, 2}	0			
1	{5, 1, 2, 3}	0			
2	{5, 1, 2, 3, 4}	∞	{4}	{3}	∅
$T \leftarrow \{4\} \cup$ QuickXplain({4, 5, 1, 2}, ∅) = {4} ∪ ∅ = {4}					
Compute QuickXplain({4, 5}, {1, 2})					
0	{4, 5}	0			
1	{4, 5, 1}	0			
2	{4, 5, 1, 2}	∞	{2}	{1}	∅
$T \leftarrow \{2\} \cup$ QuickXplain({4, 5, 2}, ∅) = {2} ∪ ∅ = {2}					

```
T ← ∅; U ← {1, ..., n}; m ← n;
while U ≠ ∅ do
    C ← T; k ← 0;
    while v*(x(C)) < v*(x̄) and k < m do
    |   k ← k + 1; C ← C ∪ {k}
    end
    if k = 0 then stop; generate cut (3.5) with x̂ = x(C);
    T ← T ∪ {k}; U ← U \ {k, ..., m}; m ← k − 1
end
```

Algorithm 5: QuickXplain without binary search, where $\bar{x} = (\bar{x}_1, \ldots, \bar{x}_n)$ with $\bar{x}_j = 1$ for $j = 1, \ldots, n$

When monotonicity is lacking. When $v^*(x)$ is not monotone, it may nonetheless be possible to strengthen cuts using the foregoing methods by repeatedly solving a problem-specific relaxation of the subproblem (rather than the original subproblem) in which some of the x_js

---

$J \leftarrow$ **QuickXplain**$(\emptyset, \{1, \ldots, n\})$;
generate cut (3.5) with $\hat{x} = x(J)$;

function QuickXplain (C, U));
if $v^*(x(C)) = v^*(\bar{x})$ **then return** $\emptyset$;
suppose U is $\{j_1, \ldots, j_m\}$;
$k \leftarrow 0$;
while $v^*(x(C)) < v^*(\bar{x})$ *and* $k < m$ **do**
  $\quad C_k \leftarrow C$;
  $\quad k \leftarrow k + 1; C \leftarrow C \cup \{j_k\}$;
end
$T \leftarrow \{j_k\}$;
split U into $U_1 = \{j_1, \ldots, j_i\}$ and $U_2 = \{j_{i+1}, \ldots, j_{k-1}\}$;
if $U_2 \neq \emptyset$ **then**
  $\quad C \leftarrow C_i \cup T$;
  $\quad T \leftarrow T \cup$ **QuickXplain**(C, U_2)
end
if $U_1 \neq \emptyset$ **then**
  $\quad C \leftarrow C_0 \cup T$;
  $\quad T \leftarrow T \cup$ **QuickXplain**(C, U_1)
end
return T;

Algorithm 6: QuickXplain with binary search, where $\bar{x} = (\bar{x}_1, \ldots, \bar{x}_n)$ with $\bar{x}_j = 1$ for $j = 1, \ldots, n$

appear as unfixed variables. The relaxation must be carefully designed so that it is tractable despite the presence of x_js, which do not appear in as variables in the original subproblem. In addition, the infeasibility of the relaxed subproblem must imply the infeasibliity of the original subproblem for any values assigned to the unfixed variables. This strategy is developed and illustrated in [149].

3.2.5 Multivalent Optimality Cuts

One drawback of the strengthened optimality cuts so far considered is that they can provide only one meaningful bound, namely the current optimal value $v^*(\bar{x})$ of the subproblem. It would be useful to have weaker but meaningful bounds that apply when x is more distant from the current master problem solution $\bar{x}$.

One can imagine several schemes for accomplishing this. A simple one is to start with a strengthened optimality cut (3.6), remove one of the variables, and write a cut based on the smaller subproblem value that results. That is, we flip a selected variable $\hat{x}_k$ to 0 to obtain $\tilde{x}$ and write the cut

$$z \geqslant v^*(\tilde{x}) - \left(v^*(\tilde{x} - \underline{v}\right) \sum_{j \in J_1(\tilde{x})} (1 - x_j)$$

One would ideally delete a variable x_k that is expected to have a minor impact on the sub-problem value. If desired, such cuts can be written for several variables x_k. In addition, any cut so obtained can be strengthened as described in Sect. 3.2.4 if the computational effort is warranted. As an example, we can start with the irreducible cut (3.7) and remove variable x_2 to obtain the cut $z \geqslant 4 - 3(1 - x_3)$, since in this case $\tilde{x} = (0, 1, 0, 0)$ and $v^*(\tilde{x}) = 4$. Removing variable x_3 yields the cut $z \geqslant 4 - 3(1 - x_2)$. These cuts cannot be usefully strengthened because they are already reduced to one variable.

One may also wish to explore the consequences of removing several variables simultaneously. The question is how to identify a promising subset of variables to remove. A heuristic used successfully in [123] begins with the optimal master problem solution $\bar{x}$ and observes which variables can be removed from the monotone cut (3.2), one at a time, without reducing the lower bound. Then the subproblem is solved with these variables all set to 0 simultaneously, whereupon the resulting optimal value serves as the lower bound in the cut. To make this more precise, it is convenient to write the optimal subproblem value $v^*(x)$ for a given x as $v^*(J)$ when $J = \{j \mid x_j = 1\}$. Now if J indexes the variables in the cut (3.2), we let

$$R = \left\{ j \in J \mid v^*(J \setminus \{j\}) = v^*(J) \right\}$$

be the index set of variables to be removed. Then we have the cut

$$z \geqslant v^*(J \setminus R) - \left(v^*(J \setminus R) - \underline{v}\right) \sum_{j \in J \setminus R} (1 - x_j)$$

which is added to the master problem alongside (3.2).

3.3   Generic Analytical Cuts

A frequently used strategy for generating analytical cuts is to determine how much the optimal subproblem value is reduced when certain tasks are removed from an assignment to some facility. If an upper bound on this reduction can be derived, it is straightforward to write a monotone optimality cut.

We again define $v^*(J)$ to be the same as $v^*(x)$ when $J = \{j \mid x_j = 1\}$. Thus in the present context, $v^*(J)$ is the optimal subproblem value when the tasks in J are assigned to a facility. We also suppose that v^* is monotone nondecreasing, because we are interested in obtaining a monotone cut.

Assuming that the tasks in J have been assigned to a facility, suppose that $v^*(J)$ satisfies the following condition:

(A1) For any set $J' \subseteq J$, removing task j from J' reduces the optimal subproblem value $v^*(J')$ by at most α_j.

Then we have the valid cut

$$z \geqslant v^*(J) - \sum_{j \in J} \alpha_j (1 - x_j) \qquad (3.9)$$

because each task j removed from J reduces the bound $v^*(J)$ by at most α_j. It is often easier to check whether (A1) is satisfied by asking whether adding a task j to J' increases $v^*(J')$ by at most α_j. Or, more precisely,

$$v^*(J' \cup \{j\}) \leqslant v^*(J') + \alpha_j, \quad \text{all } J' \subseteq J \setminus \{j\}, \text{ all } j \in J \qquad (3.10)$$

This is equivalent to the condition

$$v^*(J_1 \cup J_2) \leqslant v^*(J_1) + \sum_{j \in J_2} \alpha_j, \quad \text{all } J_1, J_2 \subseteq J$$

and therefore is somewhat akin to a submodularity property.

Condition (A1) holds in some important applications. For example, we will see in Sect. 3.4.4 that it holds for a properly defined α_j in minimum tardiness scheduling problems. Or, suppose $v^*(J)$ is the minimum number of agents required to serve the customers in J. Then adding another customer to a given set $J' \subseteq J$ of customers requires at most one more agent than required by the optimal solution for J'. We can always dedicate an extra agent to that one customer. So, (3.10) is satisfied, and we have a valid cut (3.9) with $\alpha_j = 1$ for all $j \in J$.

As another example, suppose $v^*(J)$ is the minimum travel distance required for a single vehicle to visit the cities in J, and there are no time window restrictions on visits. If a city is added to any set $J' \subset J$ of cities, the vehicle can visit that city by inserting it between any two consecutive cities on the optimal route for J'. Thus if the travel distance from city j to k is d_{jk}, inserting city j requires the vehicle to travel an additional distance of at most

$$\alpha_j = \max_{k \in J} \{d_{kj}\} + \max_{k \in J} \{d_{jk}\} \qquad (3.11)$$

We therefore have a valid cut (3.9).

However, a stronger cut is possible for vehicle routing. If multiple cities are added to J', a route connecting these cities can be inserted between any two cities in the optimal route for J', rather than inserting each city individually. The resulting cut, which does not have the generic form (3.9), is given in Theorem 3.5 of Sect. 3.5.1 below. This illustrates the fact that a generic analytical cut is often not the best available for a given application. Most of the scheduling cuts derived in the next section, for example, do not have the form of generic cuts.

One can occasionally strengthen cut (3.9) in a useful way by lifting (adding variables). Suppose $v^*(J)$ satisfies the following condition:

(A2) For any set of tasks $J' \supseteq J$, adding task j to J' increases the optimal subproblem value $v^*(J')$ by at least β_j.

Then if $v^*(J)$ satisfies (A1) as well, we have the valid cut

$$z \geqslant v^*(J) - \sum_{j \in J} \alpha_j (1 - x_j) + \sum_{j \notin J} \beta_j x_j \tag{3.12}$$

Condition (A2) can be written

$$v^*(J' \cup \{i\}) \geqslant v^*(J') + \beta_j, \quad \text{all } J' \supseteq J, \text{ all } j \notin J$$

Suppose, for example, that $v^*(J)$ is the optimal cost of visiting the cities in set J, and there are again no time window restrictions. This cost consists of the travel costs d_{jk} of moving from city j to k and fixed costs f_j of visiting any city j. Then adding a new city $j \notin J$ to any superset of the cities in J increases cost by at least f_j. Thus we set $\beta_j = f_j$ for $j \notin J$. If α_j is as given by (3.11) for $j \in J$, we have a valid cut (3.12).

3.4   Analytical Cuts for Scheduling

Analytical cuts are most highly developed for LBBD applications that have sequencing and scheduling subproblems; that is, subproblems in which the scheduling task involves determining the order in which jobs are processed. This section reviews some cut families for disjunctive and cumulative scheduling with time windows. Disjunctive scheduling requires that jobs run one at a time, while cumulative scheduling permits multiple jobs to run in parallel, subject to resource limits. Sequence-dependent setup times can incorporated into the disjunctive scheduling model if desired.

3.4.1   General Problem Structure

In a scheduling context, binary master problem variables x_j typically indicate whether job j is to be scheduled. If there are multiple facilities, the master problem contains variables x_{ij} to represent whether job j is assigned to be scheduled on facility i. The subproblem schedules the tasks after they have been assigned to facilities. We again use the notation $v^*(J)$ for the optimal value of the scheduling subproblem when the jobs in J are assigned to the facility.

We show here how to develop analytical cuts for a single facility. If there are multiple facilities that are scheduled separately, the analysis to follow can be applied to any one of these facilities. In this case, cut $z \geqslant B_{\bar{x}}(x)$ given here for a facility i will appear in the master problem as $z_i \geqslant B_{\bar{x}}^i(x)$, as explained in Sect. 2.6. The master problem combines this and similar inequalities for other facilities with an inequality of the form $z \geqslant \phi(z_1, \ldots, z_m)$, where ϕ is nondecreasing. For example, if we are minimizing total tardiness, the master problem includes the constraint $z \geqslant \sum_i z_i$, where z_i represents the tardiness incurred by facility i. If we are minimizing makespan, the master problem can omit the constraint $z \geqslant \max_i\{z_i\}$ and contain cuts of the form $z \geqslant B_{\bar{x}}^i(x)$ for each i, as illustrated in the example of Sect. 1.4.

The *disjunctive scheduling* problem asks how to schedule a set J of n jobs sequentially to minimize a stated objective, subject to time windows. Each job has processing time p_j and time window $[r_j, d_j]$ within which it should be processed. Here, r_j is a release time, and d_j is a hard (inviolable) deadline or soft due date, depending on the specific problem. The setup time (if any) for job j is σ_{kj} when it immediately follows job k. When the setup time (if any) is not sequence dependent, it is factored into the processing time p_j, and we set $\sigma_{kj} = 0$. The solution of the scheduling subproblem is a tuple of start times $s = (s_1, \ldots, s_n)$, where s_j is the start time of job j.

Cumulative scheduling is a generalization of disjunctive scheduling that allows jobs to run in parallel, so along as their total rate of resource consumption is at most C. Each job j consumes resources at a rate c_j while running. Disjunctive scheduling is a special case of cumulative scheduling in which $C = 1$ and each $c_j = 1$. The cuts developed below are valid for both disjunctive and cumulative scheduling, although we suppose each $\sigma_{kj} = 0$ for cumulative scheduling, since sequence-dependent setup times have no clear meaning in this context.

A general computational study of factors that influence LBBD performance in facility assignment and scheduling problems appears in [57].

3.4.2 Generic Scheduling Cut

As in the case of the generic analytical cuts discussed above, the central question for designing analytical cuts in a scheduling context is how much the optimal subproblem value decreases when certain jobs are removed from a facility. Thus if set J of jobs are currently assigned to the facility, and we remove some jobs to obtain a smaller set J', we wish to analyze $v^*(J) - v^*(J')$. In what follows, we will obtain bounds of the form

$$v^*(J) - v^*(J') \leqslant \sum_{j \in J \setminus J'} h_j + H \tag{3.13}$$

This gives rise to an analytical Benders cut as stated in the following lemma. A linearization is also given, since it is required in the common situation where the master problem is an MILP problem.

Lemma 3.1 *If the jobs in J are currently assigned to a facility, and inequality (3.13) holds for any $J' \subseteq J$, we have the valid analytical Benders cut*

$$z \geq \begin{cases} v^*(J) - \sum_{j \in J}(1 - x_j)h_j - H, & \text{if } x_j = 0 \text{ for some } j \in J \\ v^*(J), & \text{otherwise} \end{cases} \tag{3.14}$$

which can be linearized as follows

$$z \geq v^*(J) - \sum_{j \in J}(1 - x_j)h_j - H \tag{3.15}$$

$$z \geq v^*(J) - \sum_{j \in J}(1 - x_j)(h_j + H) \tag{3.16}$$

Proof Since job $j \in J \setminus J'$ if and only if $1 - x_j = 1$, the summation $\sum_{j \in J \setminus J'} h_j$ in (3.13) becomes $\sum_{j \in J}(1 - x_j)h_j$. Thus from (3.13), removing one or more jobs from J reduces the optimal subproblem value by at most $\sum_{j \in J}(1 - x_j)h_j + H$, and the first line of (3.14) imposes a valid bound on z. If no jobs are removed from J, the second line of (3.14) obviously imposes a valid bound. Regarding the linearization, the number $\sum_{j \in J}(1 - x_j)$ of jobs removed is either greater than 1, or at most 1. In the former case, (3.15) dominates (3.16) and imposes the same bound as (3.14). In the latter case, both (3.15) and (3.16) impose the same bound as (3.14). $\qquad\square$

3.4.3 Minimizing Makespan

We begin with the minimum makespan problem, in which each d_j is a hard deadline, and the objective is to minimize the finish time of the last job to finish. It will be convenient to use the notation $r_{\min} = \min_{j \in J}\{r_j\}$ and $r_{\max} = \max_{j \in J}\{r_j\}$, and similarly for $d_{\min}$, $d_{\max}$, and $p_{\min}$. When setup times are not sequence dependent, the problem can be stated

$$\min_{M,s} \left\{ M \,\middle|\, \begin{array}{l} M \geq s_j + p_j, \ s_j \in [r_j, d_j - p_j], \ \text{all } j \\ \displaystyle\sum_{\substack{j \\ t \in [s_j, s_j + p_j)}} c_j \leq C, \ \text{all } t \in [r_{\min}, d_{\max}] \end{array} \right\}$$

When setup times are sequence dependent, we introduce a variable y_i to denote the ith job in the scheduled sequence of jobs. The problem becomes

$$\min_{M,s,y} \left\{ M \ \middle| \ \begin{array}{l} M \geqslant s_{y_i} + \sigma_{y_{i-1}y_i} + p_{y_i}, \ \text{all } i \\ s_{y_i} \in [r_{y_i}, d_{y_i} - \sigma_{y_{i-1}y_i} - p_{y_i}], \ \text{all } i \\ (y_1, \ldots, y_n) \text{ is a permutation of } (1, \ldots, n) \\ \displaystyle\sum_{\substack{j \\ t \in [s_j, s_j + p_j)}} c_j \leqslant C, \ \text{all } t \in [r_{\min}, d_{\max}] \end{array} \right\}$$

where $y_0 = 0$ and $\sigma_{0j} = 0$ for all j.

We wish to determine how much $v^*(J)$ can change when jobs are removed from J. Thus we wish to analyze the maximum reduction $v^*(J) - v^*(J')$ in makespan when only the jobs in $J' \subset J$ are assigned.

We will rely on two related lemmas, proved in [74], each of which yields an analytical cut. It is also shown in [74] that neither cut dominates the other. Moreover, both cuts are sharp, in the sense that no tighter bound is valid for all problem instances. The cuts are similar to but stronger than cuts introduced in [75, 123, 246]. The first lemma is the following.

Lemma 3.2 *If J and J' are nonempty sets of jobs with $J' \subseteq J$, we have*

$$v^*(J) - v^*(J') \leqslant \sum_{j \in J \setminus J'} (\sigma_j^{\max} + p_j) + (r_{\max} - r_{\min} - p_{\min})^+ + d_{\max} - d_{\min} \qquad (3.17)$$

where $\sigma_j^{\max} = \max_{k \in J} \{\sigma_{kj}\}$.

Proof Suppose to the contrary that (3.17) does not hold, so that

$$v^*(J) > v^*(J') + \Delta + (r_{\max} - r_{\min} - p_{\min})^+ + d_{\max} - d_{\min} \qquad (3.18)$$

where
$$\Delta = \sum_{j \in J \setminus J'} (\sigma_j^{\max} + p_j)$$

We consider two cases.

Case 1: $v^*(J') \geqslant r_{\max}$. In this case, we can start with an optimal solution for J' and obtain a feasible solution s for J by scheduling the jobs in $J \setminus J'$ consecutively after time $v^*(J')$ in arbitrary order. To show that s is feasible, we note that its makespan M' is at most $v^*(J') + \Delta$. Then if $v^*(J') + \Delta \leqslant d_{\min}$, solution s is feasible because all deadlines are met. But this contradicts (3.18) because M' is at least as large as the optimal makespan $v^*(J)$, and the remaining terms of (3.18) have a nonnegative sum. On the other hand, if $v^*(J') + \Delta > d_{\min}$, then we have

$$v^*(J') + \Delta + d_{\max} > d_{\min} + d_{\max}$$

which implies

$$v^*(J') + \Delta + d_{\max} > d_{\min} + v^*(J)$$

because $v^*(J) \leqslant d_{\max}$. But this last inequality contradicts (3.18).

Case 2: $v^*(J') < r_{\max}$. We now create a solution s for J by scheduling the remaining jobs consecutively after $r_{\max}$. We note first that

$$(r_{\max} - r_{\min} - p_{\min})^+ = r_{\max} - r_{\min} - p_{\min}$$

due to the case hypothesis and $v^*(J') \geqslant r_{\min} + p_{\min}$. Thus (3.18) becomes

$$v^*(J) > v^*(J') + \Delta + r_{\max} - r_{\min} - p_{\min} + d_{\max} - d_{\min} \qquad (3.19)$$

We first note that if $r_{\max} + \Delta \leqslant d_{\min}$, then s is a feasible solution with makespan $M' = r_{\max} + \Delta$. This contradicts (3.19) because $v^*(J) \leqslant M'$, $v^*(J') \geqslant r_{\min} + p_{\min}$, and $d_{\max} - d_{\min} \geqslant 0$. On the other hand, if $r_{\max} + \Delta > d_{\min}$, we have

$$r_{\max} + \Delta + d_{\max} > d_{\min} + d_{\max}$$

Due to the case hypothesis, this implies

$$v^*(J) < r_{\max} + \Delta + d_{\max} - d_{\min}$$

which contradicts (3.19) because $v^*(J') \geqslant r_{\min} + p_{\min}$. The lemma follows. □

Due to Lemma 3.2, we have

Theorem 3.1 *If the jobs in set J are currently assigned to a facilty, the following inequalities comprise a valid analytical Benders cut for minimizing makespan in a disjunctive or cumulative scheduling problem:*

$$z \geqslant v^*(J) - \sum_{j \in J} (1 - x_j)(\sigma_j^{\max} + p_j) - (r_{\max} - r_{\min} - p_{\min})^+ - (d_{\max} - d_{\min})$$

$$z \geqslant v^*(J) - \sum_{j \in J} (1 - x_j)\big(\sigma_j^{\max} + p_j + (r_{\max} - r_{\min} - p_{\min})^+ + d_{\max} - d_{\min}\big)$$

$$(3.20)$$

where $\sigma_j^{\max} = \max_{k \in J} \{\sigma_{kj}\}$ and each $\sigma_{kj} = 0$ in the case of cumulative scheduling.

The second lemma deals with release times in a slightly different way. We use the notation α^+ for $\max\{0, \alpha\}$.

Lemma 3.3 *If J and J' are nonempty sets of jobs with $J' \subseteq J$, we have*

$$v^*(J) - v^*(J') \leqslant \sum_{j \in J \setminus J'} \big(\sigma_j^{\max} + p_j + (r_j - r_{\min} - p_{\min})^+\big) + d_{\max} - d_{\min}$$

where $\sigma_j^{\max} = \max_{k \in J} \{\sigma_{kj}\}$.

Proof Assume contrary to the claim that

$$v^*(J) > v^*(J') + \sum_{j \in J \setminus J'} \left(\Delta_j + (r_j - r_{\min} - p_{\min})^+ \right) + d_{\max} - d_{\min} \tag{3.21}$$

where $\Delta_j = s_j^{\max} + p_j$. If job k has the latest release time $r_k = r_{\max}$, we consider two cases based on whether job k is one of those removed from J.

Case 1: $k \in J'$. This means that $v^*(J') \geqslant r_{\max}$ as in Case 1 of the previous proof. Thus we schedule the jobs in $J \setminus J'$ consecutively after time $v^*(J')$ to obtain a solution s for J with makespan at most $v^*(J') + \sum_{j \in J \setminus J'} \Delta_j$. If we suppose that

$$v^*(J') + \sum_{j \in J \setminus J'} \Delta_j \leqslant d_{\min} \tag{3.22}$$

then solution s is clearly feasible, and we have

$$v^*(J) \leqslant v^*(J') + \sum_{j \in J \setminus J'} \Delta_j$$

But this contradicts (3.21) because $(r_j - r_{\min} - p_{\min})^+ \geqslant 0$ and $d_{\max} - d_{\min} \geqslant 0$. On the other hand, if we suppose (3.22) does not hold, we have

$$v^*(J') + \sum_{j \in J \setminus J'} \Delta_j + d_{\max} > d_{\min} + d_{\max}$$

which implies

$$v^*(J') + \sum_{j \in J \setminus J'} \Delta_j + d_{\max} > d_{\min} + v^*(J)$$

because $v^*(J) \leqslant d_{\max}$. But this implies

$$v^*(J') + \sum_{j \in J \setminus J'} \left(\Delta_j + (r_j - r_{\min} - p_{\min})^+ \right) + d_{\max} > d_{\min} + v^*(J)$$

which contradicts (3.21).

Case 2: $k \notin J'$. In this case, we have from (3.21) that

$$v^*(J) > v^*(J') + \sum_{j \in J \setminus J'} \Delta_j + \sum_{\substack{j \in J \setminus J' \\ j \neq k}} (r_j - r_{\min} - p_{\min})^+ + (r_{\max} - r_{\min} - p_{\min})^+ + d_{\max} - d_{\min}$$

which implies (3.18). But it was shown in the proof of Lemma 3.2 that (3.18) leads to a contradiction. The present lemma follows. $\square$

Due to Lemma 3.3, we have

Theorem 3.2 *If the jobs in set J are currently assigned to a facilty, the following inequalities comprise a valid analytical Benders cut for minimizing makespan in a disjunctive or cumulative scheduling problem:*

$$
\begin{aligned}
z &\geqslant v^*(J) - \sum_{j \in J} (1 - x_j)\left(\sigma_j^{\max} + p_j + (r_j - r_{\min} - p_{\min})^+\right) - (d_{\max} - d_{\min}) \\
z &\geqslant v^*(J) - \sum_{j \in J} (1 - x_j)\left(\sigma_j^{\max} + p_j + (r_j - r_{\min} - p_{\min})^+ + (d_{\max} - d_{\min})\right)
\end{aligned}
\tag{3.23}
$$

where $\sigma_j^{\max} = \max_{k \in J}\{\sigma_{kj}\}$ and each $\sigma_{kj} = 0$ in the case of cumulative scheduling.

We can compare the two cuts (3.20) and (3.23) in the example of Fig. 3.1. Here, $(r_1, r_2, r_3) = (0, 2, 2)$, $r_{\min} = 0$, $r_{\max} = 2$, $d_{\min} = d_{\max} = 6$, $(p_1, p_2, p_3) = (1, 2, 2)$ and $p_{\min} = 1$. Thus the cut (3.20) is

$$
\begin{aligned}
z &\geqslant 5 - (1 - x_1) - 2(1 - x_2) - 2(1 - x_3) \\
z &\geqslant 6 - 2(1 - x_1) - 3(1 - x_2) - 3(1 - x_3)
\end{aligned}
\tag{3.24}
$$

The cut (3.23) is

$$
\begin{aligned}
z &\geqslant 6 - (1 - x_1) - 3(1 - x_2) - 3(1 - x_3) \\
z &\geqslant 6 - 2(1 - x_1) - 3(1 - x_2) - 3(1 - x_3)
\end{aligned}
\tag{3.25}
$$

We can observe that neither cut dominates the other. When only job 1 is assigned to the facility, the two cuts yield the bounds $z \geqslant 1$ (which is sharp) and $z \geqslant 0$, respectively, so that cut (3.24) is stronger. When jobs 2 and 3 are assigned, the cuts yield the bounds $z \geqslant 4$ and $z \geqslant 5$, so that cut (3.25) is stronger.

3.4.4   Minimizing Total Tardiness

In the minimum tardiness problem, each job j has a soft due date d_j for completion, and its tardiness is how much its processing surpasses the due date. Thus, the tardiness of job j is $(s_j + p_{ij} - d_j)^+$, where s_j is the start time. The objective is to minimize the summed tardiness of all jobs. When setup times are not sequence dependent, the minimum tardiness problem can be stated

$$
\min_{s, T} \left\{ \sum_j T_j \;\middle|\; \begin{array}{l} T_j \geqslant (s_j + p_j - d_j)^+, \; s_j \geqslant r_j, \; \text{all } j \\ \displaystyle\sum_{\substack{j \\ t \in [s_j, s_j + p_j)}} c_j \leqslant C, \; \text{all } t \in [r_{\min}, r_{\max} + \textstyle\sum_j p_j] \end{array} \right\}
$$

When setup times are sequence dependent, we again let y_i be the ith job in the schedule, and the problem becomes

$$\min_{s,T} \left\{ \sum_j T_j \; \middle| \; \begin{array}{c} T_{y_i} \geq (s_{y_i} + \sigma_{y_{i-1} y_i} + p_{y_i} - d_{y_i})^+, \text{ all } i \\ s_j \geq r_j, \text{ all } j \\ (y_1, \ldots, y_n) \text{ is a permutation of } (1, \ldots, n) \\ \sum_{\substack{j \\ t \in [s_j, s_j + p_j)}} c_j \leq C, \text{ all } t \in [r_{\min}, r_{\max} + \sum_j p_j] \end{array} \right\} \tag{3.26}$$

The key lemma for writing an analytical cut is that condition (A1) is satisfied. This can be stated as follows.

Lemma 3.4 *If J and J' are sets of jobs with $J' \subseteq J$, we have*

$$v^*(J) - v^*(J') \leq \sum_{j \in J \setminus J'} \left(r_{\max} + \sum_{k \in J} (\sigma_k^{\max} + p_k) - d_j \right)^+ \tag{3.27}$$

Proof We first observe that there is a minimum tardiness solution s' for J' whose makespan M' satisfies

$$M' \leq r_{\max} + \sum_{k \in J'} (\sigma_k^{\max} + p_k) \tag{3.28}$$

This is because we may assume that the earliest job in s' starts no later than $r_{\max}$. Otherwise, we can shift all jobs to an earlier time, without increasing total tardiness, so that the first job begins at $r_{\max}$. The makespan M' of s' is no greater than the result of scheduling the jobs of J' consecutively so that the earliest job begins at $r_{\max}$. This implies (3.28). We now extend s' to obtain a solution s for J by scheduling the jobs in $J \setminus J'$ consecutively, in any order, so that the earliest job begins at $\max\{M', r_{\max}\}$. This schedule is clearly feasible because it observes all release times. The tardiness of job $j \in J \setminus J'$ in s is at most

$$\left(r_{\max} + \sum_{k \in J'} (\sigma_k^{\max} + p_k) + \sum_{k \in J \setminus J'} (\sigma_k^{\max} + p_k) - d_j \right)^+ = \left(r_{\max} + \sum_{k \in J} (\sigma_k^{\max} + p_k) - d_j \right)^+$$

because its finish time is no later than $\max\{M', r_{\max}\}$ plus the total processing time of all remaining jobs. Thus the total tardiness $\hat{T}$ of s satisfies

$$\hat{T} \leq v^*(J') + \sum_{j \in J \setminus J'} \left(r_{\max} + \sum_{k \in J} (\sigma_k^{\max} + p_k) - d_j \right)^+ + v^*(J')$$

Since $v^*(J) \leq \hat{T}$ due to the optimality of $v^*(J)$, we have (3.27). □

Lemma 3.4 implies that we have an analytical cut (3.9) for the minimum tardiness problem, where

$$\alpha_j = \left(r_{\max} + \sum_{k \in J} (\sigma_k^{\max} + p_k) - d_j \right)^+, \text{ all } j \in J$$

Fig. 3.2 A minimum tardiness
schedule

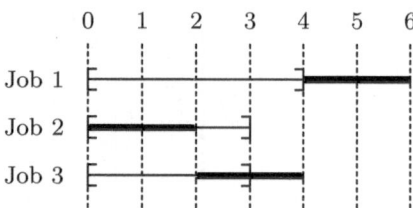

This proves the following.

Theorem 3.3 *If the jobs in set J are currently assigned to a facility, the following inequality is a valid analytical Benders cut for minimizing total tardiness in a disjunctive or cumulative scheduling problem:*

$$z \geqslant v^*(J) - \sum_{j \in J}(1 - x_j)\left(r_{\max} + \sum_{k \in J}(\sigma_k^{\max} + p_k) - d_j\right)^+ \tag{3.29}$$

where $\sigma_k^{\max} = \max_{\ell \in J}\{\sigma_{\ell k}\}$ and each $\sigma_{\ell k} = 0$ in the case of cumulative scheduling.

As an example, consider the disjunctive scheduling problem depicted in Fig. 3.2. Here $J = \{1, 2, 3\}$, $r_{\max} = 0$, $p = (2, 2, 2)$, and $d = (4, 3, 3)$. A minimum tardiness solution is shown, with total tardiness $v^*(J) = 3$. The cut (3.29) is

$$z \geqslant 3 - 2(1 - x_1) - 3(1 - x_2) - 3(1 - x_3)$$

Aside from the obvious tardiness bound of 3 for $x = (1, 1, 1)$ (no jobs are removed), the cut obtains a sharp bound of 1 for $x = (0, 1, 1)$ (job 1 is removed). It provides no positive bound for other values of x, but no positive bound is valid, because tardiness drops to zero in these cases.

3.4.5 Minimizing Number of Late Jobs

A subproblem that minimizes the number of late jobs can generate either feasibility cuts or optimality cuts. If all jobs must finish on time, the scheduling subproblem can be viewed as infeasible when the number of late jobs cannot be reduced to zero, thus giving rise to a feasiblity cut. If late jobs can be tolerated, the subproblem generates an optimality cut.

We first consider feasibility cuts. As before, let J be the set of jobs assigned to a facility. If not all jobs in J can be feasibly scheduled, we can generate an analytical feasibility cut if we know which jobs are late in the optimal solution. If $\bar{J}$ is the set of late jobs, then $|\bar{J}|$ is the minimum number of late jobs in J. Now if any one of the jobs in $\bar{J}$ is assigned to the facility, at least one of the on-time jobs must be removed, since otherwise $|\bar{J}|$ is not the minimum number of late jobs in J. This yields a feasibility cut consisting of the following inequalities

$$(1 - x_j) + \sum_{k \in J \setminus \bar{J}} (1 - x_i) \geqslant 1, \quad \text{all } j \in \bar{J} \tag{3.30}$$

A slight extension of this reasoning allows us to derive an optimality cut. It states that if ℓ currently late jobs are assigned to a facility, and all currently on-time jobs are assigned, then there are at least ℓ late jobs in the resulting schedule.

Theorem 3.4 *Let J be the set of jobs assigned to a facility. Given any solution of a disjunctive or cumulative scheduling problem on the facility that minimizes the number of late jobs, let $\bar{J}$ be the set of late jobs in the solution. Then a valid analytical Benders cut is*

$$z \geqslant \sum_{j \in \bar{J}} x_j - |\bar{J}| \sum_{j \in J \setminus \bar{J}} (1 - x_j)$$

Proof Suppose that the jobs in set K are assigned to a facility, where K contains a nonempty subset L of $\bar{J}$ and all jobs in $J \setminus \bar{J}$, possibly among other jobs. It suffices to show that at least $|L|$ jobs are late in any schedule for K. Assume to the contrary that fewer than $|L|$ jobs are late in some schedule for K. Add the jobs in $\bar{J} \setminus L$ to this schedule in such a way that they are processed after the jobs in K and are all late. This yields a schedule that contains all the jobs in $K \cup J$ and fewer than $\bar{J}$ late jobs. Then if we remove the jobs not in K from this schedule, we obtain a schedule of the jobs in J in which fewer than $|\bar{J}|$ are late. This contradicts the assumption that there are a minimum of $|\bar{J}|$ late jobs in any schedule for J. $\square$

3.5   Analytical Cuts for Vehicle Routing

We develop analytical cuts for two types of vehicle routing problems. One is a basic multiple-vehicle problem in which the master problem assigns locations that each vehicle must visit, the subproblem decouples into a routing problem for each vehicle. The other is a single-vehicle problem with time windows. For this problem, we assume a decomposition in which the master problem determines the sequence in which the vehicle visits locations, and the subproblem seeks a travel schedule that satisfies the time windows.

3.5.1   Optimality Cuts for Multiple Vehicle Routing

Perhaps the best known vehicle routing problem is one in which there are several vehicles, and each is assigned a subset of customers to visit. The customers are located on a transportation network in which each arc is associated with a travel distance, travel time, or some other cost measure. Each vehicle is routed to visit its customers in an order that minimizes total cost. In a typical Benders approach, the master problem assigns customers to vehicles, and the subproblem decouples into a separate routing problem for each vehicle.

If a vehicle's trip can start at any customer location and end at any customer location, the routing problem is a special case of the minimum makespan disjunctive scheduling problem of Sect. 3.4.3. Setup times σ_{jk} become arc costs, processing times p_j are all zero, and there are no time windows ($r_j = 0$ and $d_j = \infty$ for all j). The analytical cuts given in Sect. 3.4.3 can therefore be specialized to the routing problem. However, in most applications, the vehicles must start and end at a depot, which makes cut generation slightly different.

We suppose that the master problem has assigned the customers in set $J = \{1, \ldots, m\}$ to a given vehicle. The routing problem is to find a permutation $y_1, \ldots, y_m$ of $1, \ldots, m$ that minimizes the cost of a tour that begins at the depot (customer 0), visits $y_1, \ldots, y_m$ in that order, and returns to the depot. If c_{jk} is the cost of traversing arc (j, k), the cost of a tour $\boldsymbol{y}$ is

$$v(\boldsymbol{y}) = c_{0y_1} + \sum_{i=1}^{n-1} c_{y_i y_{i+1}} + c_{y_m 0} \tag{3.31}$$

If an optimal tour has cost $v^*(J)$, we wish to write a valid cut that bounds the cost when some customers are removed from the vehicle's assignment. This is accomplished by providing a simple analog of Lemma 3.2.

Lemma 3.5 *If J and J' are nonempty sets of customers with $J' \subseteq J$, we have*

$$v^*(J) - v^*(J') \leqslant \sum_{j \in J \setminus J'} c_j^{\max} + c_0^{\max}$$

where $c_j^{\max} = \max\{c_{kj} \mid k \in J \cup \{0\}\}$.

Proof Suppose the customers are indexed so that $J' = \{1, \ldots, m'\}$, and suppose $(0, \boldsymbol{y}^1, 0)$ is an optimal tour for the customers in J' with cost $v(\boldsymbol{y}^1) = v^*(J')$, where $\boldsymbol{y}^1 = (y_1, \ldots, y_{m'})$. Let $\boldsymbol{y}^2 = (y_{m'+1}, \ldots, y_m)$ be an arbitrary path connecting the customers in $J \setminus J' = \{m' + 1, \ldots, m\}$. Then $(0, \boldsymbol{y}^1, \boldsymbol{y}^2, 0)$ is a tour for all the customers in J with cost $v(\boldsymbol{y}^1, \boldsymbol{y}^2)$. Since $v^*(J) \leqslant v(\boldsymbol{y}^1, \boldsymbol{y}^2)$, it suffices to show

$$v(\boldsymbol{y}^1, \boldsymbol{y}^2) - v^*(J') \leqslant \sum_{j \in J \setminus J'} c_j^{\max} + c_0^{\max} \tag{3.32}$$

But we have from (3.31) that

$$v(\boldsymbol{y}^1, \boldsymbol{y}^2) - v(\boldsymbol{y}^1) = -c_{y_{m'} 0} + \sum_{i=m'}^{m-1} c_{y_i, y_{i+1}} + c_{y_m 0}$$

This implies (3.32) because $v(y^1) = v^*(J')$, $c_{y_m 0} \leqslant c_0^{max}$, and $c_{y_i, y_{i+1}} \leqslant c_{y_{i+1}}^{max}$ for $i = m', \ldots, m - 1$, due to the definition of c_j^{max}. $\qquad \square$

Due to Lemma 3.1, we can infer from Lemma 3.2 a valid optimality cut.

Theorem 3.5 *If the customers in set J are currently assigned to a given vehicle, the following inequalities comprise a valid analytical Benders cut for minimizing cost in a traveling salesman problem:*

$$z \geqslant v^*(J) - \sum_{j \in J}(1 - x_j)c_j^{max} - c_0^{max}$$

$$z \geqslant v^*(J) - \sum_{j \in J}(1 - x_j)(c_j^{max} + c_0^{max})$$

where $c_j^{max} = \max_{k \in J}\{c_{kj}\}$.

3.5.2 Feasibility Cuts for Routes with Time Windows

In some vehicle routing applications, the master problem determines the route that each vehicle takes through a network, and the subproblem determines how the stops along this route can be scheduled subject to time windows. If the stops cannot be scheduled without violating time windows, a feasibility cut must be generated. A simple nogood cut would require that at least one arc on the vehicle's assigned route be removed. The cut can be strengthened heuristically by re-solving the subproblem a few times to find infeasible subpaths. However, a bit of analysis allows one to find all irreducibly infeasible subpaths with a limited number of re-solves.

Suppose that $x_{jj'} = 1$ indicates that the path assigned to given vehicle contains an arc (j, j') that connects stop j with stop j'. Suppose also that the master problem solution assigns a path to the vehicle consisting of arcs $(1, 2), (2, 3), \ldots, (m-1, m)$. The travel time on arc $(j, j+1)$ is p_j, and each stop j has time window $[r_j, d_j]$. The latter means that the vehicle must arrive no later than deadline d_j, while it can arrive before release time r_j but cannot depart before r_j. A simple greedy algorithm finds a feasible schedule, if one exists, by scheduling each stop as early as possible. Schedule stop 1 at time $s_1 = r_1$, and schedule stop j at time $s_j = \max\{s_{j-1} + p_{j-1}, r_j\}$ for $j = 2, \ldots, n$. The schedule is feasible if and only if $s_{j-1} + p_{j-1} \leqslant d_j$ for $j = 2, \ldots, m$.

When the greedy schedule is infeasible, we wish to find all irreducibly infeasible subpaths and generate a cut for each. This is accomplished as follows. Execute the greedy algorithm until it encounters the first time window violation, and suppose this occurs at stop k. Then we know that arc $(k-1, k)$ cannot be deleted from the subpath $1, \ldots, k$ while preserving infeasibility. While executing the greedy heuristic, let stop j be the latest stop with slack; that is, the latest stop for which $s_{j-1} + p_{j-1} \leqslant r_j$ (with $j = 1$ if there is no such stop).

Then arcs on the subpath from stop 1 to stop j can be deleted while preserving infeasibility. At this point we test whether arc $(j, j+1)$ can be deleted by starting the greedy heuristic at stop $j+1$ with $s_{j+1} = r_{j+1}$. Let's say that the heuristic finds the first time window violation at stop k', and stop j' is the latest stop with slack. If $k' = k$, we still have infeasiblity at stop k, and arc $(j, j+1)$ can be deleted. Otherwise, we conclude that $j, \ldots, k$ is an irreducibly infeasible subpath, because neither arc $(j, j+1)$ nor arc $(k-1, k)$ can be deleted while preserving infeasibility. This yields the cut

$$\sum_{\ell=j}^{k-1}(1 - x_{\ell,\ell+1}) \geqslant 1 \tag{3.33}$$

Now we update (j, k) to (j', k'), test as before whether arc $(j, j+1)$ can be deleted, and continue until all stops are scheduled. A formal statement of the procedure appears as Algorithm 7. The function greedy(t) applies the greedy algorithm starting at stop $t = 1$ or $t = j + 1$ and returns (j', k'). If the entire path beginning at t is feasible, the function returns $(j', m + 1)$.

Figure 3.3 illustrates the procedure on an infeasible path with 8 stops. The time window for each stop is shown at the top of the figure. The five horizontal lines below the time windows correspond to applications of the greedy algorithm. Every arc incurs a travel time of 10. The small circles indicate arrivals or departures at stops, and filled-in circles are late arrivals. Note that stop 2 is slack in the first greedy solution, as is stop 6 in the fourth solution.

$(j, k) \leftarrow$ **greedy**(1);
while $k \leqslant m$ **do**
 $\quad (j', k') \leftarrow$ **greedy**$(j + 1)$;
 $\quad$ **if** $k' = k$ **then** $j \leftarrow j + 1$;
 $\quad$ **else**
 $\quad\quad$ generate cut (3.33);
 $\quad\quad j \leftarrow j'; k \leftarrow k'$
 $\quad$ **end**
end

function greedy (t)
$j \leftarrow t; \ell \leftarrow j; s_\ell \leftarrow r_\ell$;
while $\ell \leqslant m$ **do**
 $\quad$ **if** $s_\ell + p_\ell \leqslant r_{\ell+1}$ **then** $\ell \leftarrow \ell + 1; j \leftarrow \ell; s_\ell \leftarrow r_\ell$;
 $\quad$ **else if** $s_\ell + p_\ell > d_{\ell+1}$ **then return** $(j, \ell + 1)$;
 $\quad$ **else** $\ell \leftarrow \ell + 1; s_\ell \leftarrow s_{\ell-1} + p_{\ell-1}$;
end
return $(j, m + 1)$;

Algorithm 7: Procedure for generating cuts for irreducibly infeasible paths

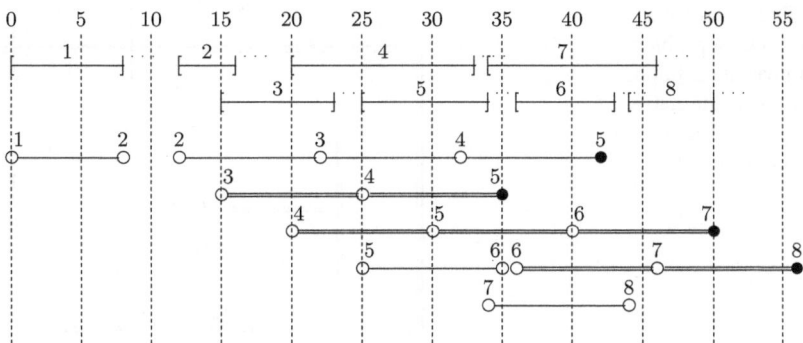

Fig. 3.3 Example of procedure for finding irreducibly infeasible subpaths (double lines). The horizontal dimension is time

The three double lines indicate irreducibly infeasible subpaths, corresponding respectively to the following cuts:

$$(1 - x_{34}) + (1 - x_{34}) \geqslant 1$$
$$(1 - x_{45}) + (1 - x_{56}) + (1 - x_{67}) \geqslant 1$$
$$(1 - x_{67}) + (1 - x_{78}) \geqslant 1$$

3.6 Analytical Cuts for Packing Problems

Space packing subproblems arise frequently and provide an opportunity to identify interesting analytical Benders cuts. As a generic packing problem, we suppose that rectangles of different sizes are to be packed into a larger rectangular space. The rectangles cannot overlap, nor can they be rotated before placement. An example adapted from [233] appears in Fig. 3.4a, which shows an infeasible packing.

We present an LBBD approach similar to that in [62]. The packing problem can be decomposed by letting the master problem determine the x-coordinates of the n rectangles to be packed, leaving the subproblem to check whether there are y-coordinates that result in a feasible packing. We assume the x-coordinates are nonnegative integers and let the master problem contain binary variables x_{jk} that take the value 1 when the left side of rectangle j has x-coordinate k. It will be convenient to say that $x_j = k$ when $x_{jk} = 1$. Then if the solution $\bar{x}$ of the master problem defines an infeasible subproblem, we have a valid monotone feasbility cut

$$\sum_{j=1}^{n} x_{j\bar{x}_j} \leqslant n - 1 \tag{3.34}$$

Fig. 3.4 Example of **a** a
rectangle packing problem,
with an infeasible soluion
shown, and **b** a cumulative
scheduling relaxation of (**a**),
with a solution shown

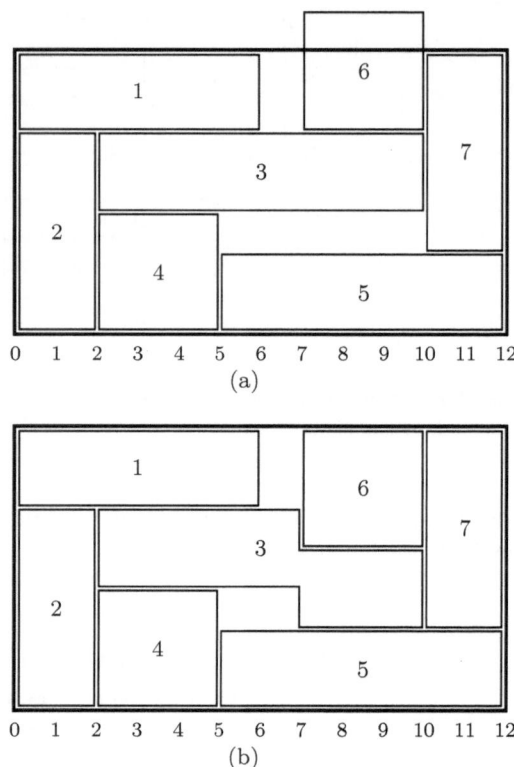

since at least one of the rectangles must be assigned a different x-coordinate. In the example
of Fig. 3.4a, the x coordinates shown create an infeasible subproblem, and we have the
simple nogood cut

$$x_{10} + x_{20} + x_{32} + x_{42} + x_{55} + x_{67} + x_{7,10} \leqslant 6 \qquad (3.35)$$

A natural approach to obtaining x-coordinates in the master problem, used in [62], is
to solve a cumulative scheduling problem (Sect. 3.4.1) that relaxes the packing problem.
Each rectangle is viewed a job whose processing time is the width w_j of the rectangle, and
whose resource cumsumption rate is the height of the rectangle. The time horizon is the
width of the space to be packed, and the resource limit is its height. The resulting start time
of each job j becomes the x-coordinate x_j of its left side. The example of Fig. 3.4a presents
a cumulative scheduling problem with the feasible solution shown in Fig. 3.4b. As already
noted, the resulting x-coordinates do not permit a feasible packing.

As a first step toward strengthening the simple nogood cut (3.34), we can employ a
bisection search to determine whether the subproblem can be decoupled. Given a solution $\bar{x}$
of the master problem for which the subproblem is infeasible, we check whether the space

can be divided by a vertical line so that every rectangle is placed wholly on one side or wholly on the other. That is, we check whether there is an x-coordinate d such that for every rectangle j, either $\bar{x}_j + w_j \leqslant d$ or $\bar{x}_j \geqslant d$. This splits the problem into two independent packing problems. One (or both) of these problems will be infeasible, and we generate a feasibility cut with fewer variables for that problem (or for both problems).

Let's suppose we arrive at a packing problem on rectangles $1, \ldots, n$ that cannot be further divided. This cut can be strengthened heuristically, as in [62], by eliminating rectangles from left to right until feasibility is obtained. Let $J_k = \{j \mid \bar{x}_j = k\}$ be the set of rectangles with x-coordinate k. We first drop from the problem all rectangles in J_k for $k = 0$. If the resulting subproblem is feasible, the procedure stops. Otherwise, we remove the variables x_{jk} from (3.34) for all $j \in J_k$ and reduce the right-hand side by $|J_k|$. We continue in this fashion for $k = 1, \ldots, n$. Alternatively, we could apply the deletion filter (Algorithm 2) to (3.34), which obtains an irreducible cut but generally requires more computation.

It may also be possible to lift cut (3.34), as noted in [62], by inserting additional variables without increasing the right-hand side. For example, we note that rectangle 1 horizontally overlaps rectangles 2, 3, 4, and 5. That is, the interval spanned by rectangle 1 on the x-axis has an overlap of positive width with each of the intervals spanned by the other rectangles. The same would be true of we shifted rectangle 1 one unit to the right, from x-coordinate 0 to 1. In general, we let $J_j(k)$ be the set of rectangles that horizontally overlap rectangle j when rectangle j has x-coordinate k, so that $J_1(0) = J_1(1) = \{1, 2, 3, 4, 5\}$. If we shift rectangle 1 one unit to the right, the infeasiblity must remain, which means that its position can be neither 0 nor 1 if no other rectangles are shifted. Similarly, rectangle 6 can be shifted a unit to the left, from position 7 to position 6, without relieving the infeasibility, because $J_6(6) = J_6(7) = \{3, 5\}$. We can therefore add variables x_{11} and x_{66} to (3.35) to obtain a lifted (and strengthened) cut

$$x_{10} + x_{11} + x_{20} + x_{32} + x_{42} + x_{55} + x_{66} + x_{67} + x_{7,10} \leqslant 6$$

In general, for each variable $x_{j\bar{x}_j}$ in cut (3.34), we can add a variable x_{jk} to the cut for each $k \neq \bar{x}_j$ for which $J_j(k) = J_j(\bar{x}_j)$.

3.7   Symmetry Cuts

A Benders cut that is found to be valid can frequently give rise to similar cuts based on symmetries in the problem. For example, if a cut prevents a certain set of tasks from being assigned to an agent, their assignment to another agent may be infeasible for the same reason. Since generation of such symmetry cuts may require much less computation than deriving them directly from the subproblem, there can be significant computational benefit in taking advantage of symmetry.

To illustrate the idea, let the binary variable $x_{ij} = 1$ when job j is assigned to machine i, and suppose the valid monotone feasiblity cut

$$\sum_{j \in J}(1 - x_{ij}) \geqslant 1 \tag{3.36}$$

has been generated for a particular machine i. Then if the machines have identical physical characteristics, we can substitute any other machine i' for i in (3.36) and obtain another valid cut. Or if the machines differ only in that some run more slowly than others, we can substitute for i any machine i' that runs at least as slow as i on any job in J. Alternatively, we might suppose $x_{ij} = 1$ when passenger j is assigned to bus i. Then given a valid cut (3.36) for a particular bus i, we can substitute for i any bus with no greater passenger capacity, provided the buses are otherwise identical.

The same principle applies to optimality cuts. If we have a valid optimality cut

$$z_i \geqslant v_i^*(J) - \sum_{j \in J} \alpha_{ij}(1 - x_j)$$

for a particular machine i, then we can substitute for i any machine i' that is identical to machine i with respect to both physical characteristics and costs it incurs.

Nogood cuts can occasionally give rise to cardinality constraints. For example, if all the tasks in a superset J' of J are identical, then a valid feasibility cut (3.36) can be replaced by the more general cut

$$\sum_{j \in J'}(1 - x_{ij}) \geqslant |J'| - |J| + 1$$

This cut subsumes all cuts obtained from (3.36) by replacing any subset of tasks in J with identical tasks.

When there is substantial symmetry in the problem, a given cut may spawn an impracticably large number of symmetry cuts. In such cases, one can investigate possibilities for reducing symmetry before cuts are generated. For example, symmetry breaking constraints might be added to the master problem formulation [91, 102], or if the problem is solved by branch and check, one might select a solver with symmetry breaking features in the branching mechanism [178]. Either option may largely obviate the need to generate symmetry cuts.

3.8   Explanation-Based and Automatic Cut Generation

The logic-based cuts so far described do not rely on direct knowledge of the inference dual solution. Rather, the cuts are based on exploratory probing that repeatedly re-solves the subproblem, or on the current solution and an analysis of problem structure.

Nonetheless, solvers are beginning to supply "explanations" of their solutions that reveal some details about their infeasiblity or optimality proof. The explanation usually takes the form of nogoods that identify premises of the proof, particularly in constraint programming and satisfiability solvers. This opens the door to cuts that are based more directly on the

Fig. 3.5 An infeasible schedule

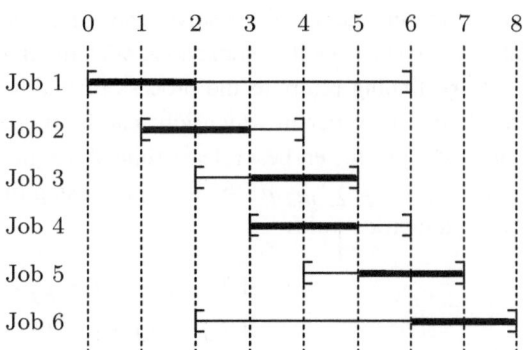

dual solution. It also suggests a pathway to automatic generation of logic-based cuts, since suitably designed general-purpose solvers can provide nogoods without user intervention. A general discussion of using explanation to generate cuts can be found in [41–43].

At this writing, automatic cut generation has been implemented for the traveling salesman problem with time windows [155], as well as in a more general form in the MiniZinc modeling language [68] and the branch-and-check solver Nutmeg [154]. These are discussed further in Sect. 4.8. All three solve the subproblem with constraint programming solvers that can generate nogood cuts in a manner similar to conflict-directed clause learning in satisfiability solvers. Since the precise method depends on the solver, we illustrate the basic ideas by example rather than attempt to develop them formally.

Consider the disjunctive scheduling subproblem depicted by Fig. 3.5. This is a pure feasiblity problem with time windows, again illustrated by the horizontal brackets. The heavy lines show the processing times but do not illustrate a feasible solution due to overlapping jobs. In fact, there is no feasible solution. The infeasiblity results from a master problem solution that assigns jobs 1–6 to this facility. Thus we have $\bar{x}_j = 1$ for $j = 1, \ldots, 6$, where x_i represents the facility assigned to job j as before.

A simple device for obtaining cuts from the solution process is to augment the subproblem with variables x'_j that mirror the master problem variables, along with constraints $x'_j = \bar{x}_j$ that assign to these variables the current master problem solution. Then when the solver generates nogoods, those nogoods that contain only the variables x'_j can be used as Benders cuts. We will examine nogoods that might be generated by a constraint programming solver.

Constraint programming solvers generally address disjunctive scheduling by applying *edge-finding* and related rules for reducing or "filtering" variable domains, such as the domains of integer-valued start-time variables s_j. This may prove infeasibility by reducing a domain to the empty set, or find a feasible solution by reducing all domains to a singleton. If both fail, branching takes over, and filtering is applied at each node of the search tree. The process continues until infeasiblility is proved or a solution found. As it happens, a single edge-finding rule can prove infeasibility of the problem in Fig. 3.5 without branching.

By examining the proof, the solver can build *conflict graphs* similar to those used in satisfiability solvers as well as some constraint and integer programming solvers.

Edge finding refers to the process of finding arcs in a precedence graph and thereby restricting the order in which jobs can be processed. Given a subset J if jobs, let $E_J = \min_{j \in J}\{r_j\}$ be the earliest release time of the jobs in J, $L_J = \max_{j \in J}\{d_j\}$ the latest release time, and $p_J = \sum_{j \in J} p_j$ the total processing time. The edge finding principle consists of two symmetrical rules

$$\text{If } L_{J \cup \{j\}} - E_J < p_j + p_J, \text{ then } j \prec J \tag{3.37}$$

$$\text{If } L_J - E_{J \cup \{j\}} < p_j + p_J, \text{ then } j \succ J \tag{3.38}$$

where $j \prec J$ means that job j must precede all the jobs in J, and the reverse for $j \succ J$.

Edge finding proves infeasibility of the example in at least two ways. First, we let $J = \{2, 3\}$ and note that rule (3.37) implies that job 1 must precede jobs 2 and 3. If $J = \{3\}$, the rule implies that job 2 must precede job 3. These two precedences force job 3 to start no earlier than time $s_3 = 4$. But since the original domain of s_3 was $\{2, 3\}$ due to its time window, the domain is reduced to the empty set and infeasibility proved. A symmetrical application of rule (3.38) reduces the domain of s_4 to the empty set and again proves infeasiblity.

The first proof of infeasibility is represented in the *conflict graph* of Fig. 3.6, and a similar graph represents the second proof. Each node corresponds to a constraint, and the arrows entering a node indicate other constraints from which it is deduced in one step of the proof. The original constraints and variable domains of the subproblem are not shown (other than the assignments $x'_j = \bar{x}_j$) because they will not appear in nogoods. Nodes with no incoming arrows are premises, and the conflict node $\emptyset$ represents a contradiction. Since the contradiction is proved, the problem is infeasible. Note that the edge-finding-based derivation of the precedence constraints is based on the premises that jobs 1, 2, and 3 are assigned to the facility.

Nogoods can now be found by partitioning the graph in a way that separates premises from the conflict node; two possible partitions are shown. The left side of the partition is the "reason side," and the right side is the "conflict side." A nogood is now formed from the nodes on the "frontier" of the reason side; that is, the set of nodes on the reason side from

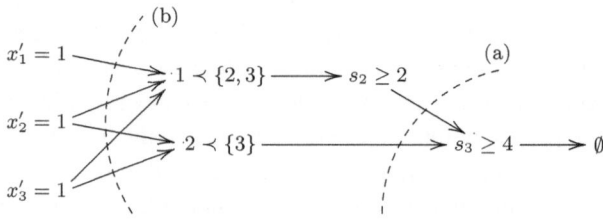

Fig. 3.6 A conflict graph with two possible partitions (**a**) and (**b**)

Fig. 3.7 A small branching tree for a disjunctive scheduling problem

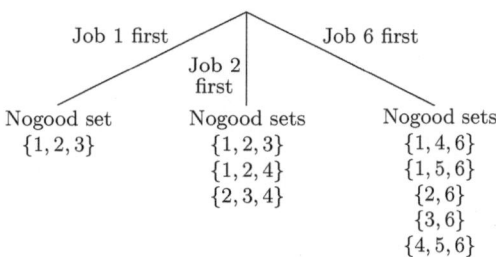

which at least one outgoing edge crosses to the other side. This is because at least one of the corresponding constraints must fail to hold if the proof of infeasibility is to be avoided. Partition (a) yields the nogood

$$(s_2 < 2) \text{ or } (3 \prec \{2\})$$

which does not provide a Benders cut because it does not contain variables x'_j. Partition (b) is more useful because it says that

$$(x'_1 = 0) \text{ or } (x'_2 = 0) \text{ or } (x'_3 = 0)$$

which yields the Benders cut $(1 - x_1) + (1 - x_2) + (1 - x_3) \geqslant 1$. In general, we cut the graph so that the reason side consists of constraints $x'_j = \bar{x}_j$. The second proof of infeasibility similarly yields the cut $(1 - x_4) + (1 - x_5) + (1 - x_6) \geqslant 1$. Both cuts can be added to the master problem.

If edge finding and similar rules do not resolve the issue of feasibility, we must branch. A popular scheme is to branch on which job is scheduled first, among those jobs not yet sequenced. To illustrate this using the same example, suppose we use a weak form of edge finding that only seeks to order pairs of jobs (i.e., it considers only singletons J). We also infer infeasibility when a job runs too long to fit in its own (reduced) window.

At the root node of the branching tree (Fig. 3.7), weakened edge finding deduces $2 \prec 3 \prec 4 \prec 5$. This implies that only job 1, 2, or 6 can be first, and we branch accordingly. At it happens, each of these branches allows us to deduce infeasibility without further branching. There is one conflict graph at the leftmost node, and it identifies nogood set $\{1, 2, 3\}$: assigning these jobs to the facility allows the weak form of edge finding to deduce infeasibility on the assumption that job 1 is first. At the middle node, three conflict graphs identify three nogood sets on the assumption that job 2 is first. The rightmost node produces five nogood sets, as shown. To derive a strong Benders cut, we wish to identify a small set of jobs that, if scheduled, create infeasibility at every leaf node of the tree. In this case, the smallest set is $\{1, 2, 3, 6\}$, which yields the Benders cut

$$(1 - x_1) + (1 - x_2) + (1 - x_3) + (1 - x_6) \geqslant 1$$

It is somewhat weaker that the cuts derived earlier, due to the weaker filtering mechanism.

Variations and Special Cases of LBBD

4

4.1 Introduction

Benders decomposition rests on a general problem-solving idea that can take several forms, one of which is the standard LBBD algorithm developed in previous chapters. Another very useful form is branch and check, which uses the same logic-based cuts but generates them *during* the solution of the master problem. A significant fraction of applications find it preferable to standard LBBD, since the master problem need only be solved once. The chapter begins with a development and example of branch and check.

In another variation of LBBD, the master problem *enumerates* solutions rather than maintaining a lower bound on the optimal solution value. We refer to this approach as enumerative LBBD. It is has occasionally been found convenient, but it places limitations on the extent to which the subproblem can decouple. It is described here, along with an example, to clarify its limitations and its differences from other forms of LBBD.

Combinatorial Benders cuts are logic-based feasiblity cuts used in the context of a specialized branch-and-check method. The master problem is a mixed integer/linear programming problem and the subproblem a linear programming problem. As this is a rather widely applied special case of LBBD, due to the prevalence of MILP in applications, we provide an exposition here.

Stochastic and robust optimization warrant special mention in the chapter as application domains for LBBD, because of their practical significance and the ability of LBBD to deal with the large number of scenarios that characterize these problems. The chapter then proceeds to three enhancements of LBBD that can improve performance. The most important is the use of a subproblem relaxation in the master problem, and two methods for doing so are presented. The others are multilevel LBBD and dynamic variable partitioning, the latter of which is used in propositional satisfiability (SAT) solvers, which can be viewed as implementing a special case of branch and check.

© The Author(s), under exclusive license to Springer Nature Switzerland AG 2024   61
J. Hooker, *Logic-Based Benders Decomposition*, Synthesis Lectures on Operations
Research and Applications, https://doi.org/10.1007/978-3-031-45039-6_4

The chapter concludes with some observations on how LBBD might be automated in an off-the-shelf solution method, and the pros and cons of doing so.

4.2   Branch and Check

Branch and check is a variation of LBBD that solves the master problem only once [122, 244]. It can be an attractive alternative when the master problem is significantly harder to solve than the subproblem and is solved by a branching procedure. When a feasible solution of the master problem is encountered during the branching process, it is "checked" by solving the subproblem that results. For example, an integral solution encountered during a branch-and-cut algorithm might be checked in this way. If the subproblem is infeasible, a feasibility cut is added to the master problem and enforced during the remainder of the tree search. The algorithm terminates when the search is exhaustive.

Branch and check is normally applied when the master problem is a mixed integer/linear programming (MILP) problem. The original problem therefore has the form

$$\min_{x,y,w} \left\{ f(x, y) \mid C(x, y), \ Ax + Dw \geq b, \ x \in \mathbb{Z}^n, \ y \in \mathcal{D}_y, \ w \in \mathbb{R}^p \right\} \tag{4.1}$$

The subproblem SP(x) for $x = x^k$ is

$$\min_y \left\{ f(x^k, y) \mid C(x^k, y), \ y \in \mathcal{D}_y \right\}$$

and its optimal value is $v^*(x^k)$. The master problem after the solution of k subproblems is the MILP problem

$$\min_{z,x,w} \left\{ z \mid Ax + Dw \geq b; \ z \geq B_{x^\ell}(x), \ \ell = 1, \ldots, k; \ x \in \mathbb{Z}^n, \ w \in \mathbb{R}^p \right\} \tag{4.2}$$

where $z \geq B_{x^\ell}(x)$ for $\ell = 1, \ldots, k$ are Benders cuts in the form of linear inequalities generated generated so far.

Let (z^k, x^k, w^k) be an optimal solution of the LP relaxation of the master problem at a given node of the branching tree, which is formed by replacing $x \in \mathbb{Z}^n$ with $x \in \mathbb{R}^n$ in (4.2). If x^k is integral, the corresponding subproblem SP (x^k) is solved. If $z_k \geq v^*(x^k)$, the tree search continues in the normal fashion. Otherwise, a Benders cut $z \geq B_{x^k}(x)$ is generated. In practice, a feasibility cut $F_{x^k}(x)$ is generated, rather than $z \geq B_{x^k}(x)$, if the subproblem is infeasible. The LP solution (z_k, x^k, w^k) is no longer feasible because it violates the cut. The branching process continues with the new cut added to the master problem, and the search must at some point return to the infeasible node and check it again. At termination, the incumbent solution (if any) is optimal for (2.6) because it defines a feasible subproblem.

Algorithm 1 specifies a branch-and-check method when the master problem is solved by a generic branch-and-bound search. The set $\mathcal{O}$ consists of currently open nodes in the search tree (nodes that have not yet been processed). A node is processed by solving the LP relaxation, and (a) if the solution is integer, solving the subproblem (and generating a cut if necessary), or (b) if the solution is noninteger, pruning the tree or branching on a fractional variable. The node must be processed again at some point (perhaps immediately) if a cut is generated in case (a). The incumbent solution is $(\boldsymbol{x}^*, \boldsymbol{y}^*, \boldsymbol{w}^*)$ and has optimal value z^*.

---

let $\mathcal{O} \leftarrow \{N_0\}$, where N_0 is the root node of the branching tree;
$z^* \leftarrow \infty$; $k \leftarrow 0$;
while $\mathcal{O}$ *is nonempty* **do**
 $\quad k \leftarrow k + 1$;
 $\quad$ select a node $N \in \mathcal{O}$ and remove it from $\mathcal{O}$;
 $\quad$ solve the LP relaxation of the problem at node N and let z_k be its optimal value;
 $\quad$ **if** $z_k < \infty$ *(the LP is feasible)* **then**
 $\quad\quad$ let $(\boldsymbol{x}^k, \boldsymbol{w}^k)$ be an optimal solution of the LP relaxation;
 $\quad\quad$ **if** $\boldsymbol{x}^k$ *is integer* **then**
 $\quad\quad\quad$ solve the subproblem $\mathrm{SP}(\boldsymbol{x}^k)$ and let $v^*(\boldsymbol{x}^k)$ be its optimal value;
 $\quad\quad\quad$ **if** $z_k < v^*(\boldsymbol{x}^k)$ **then**
 $\quad\quad\quad\quad$ generate a Benders cut $z \geq B_{\boldsymbol{x}^k}(\boldsymbol{x})$ & add it to the master problem (4.2);
 $\quad\quad\quad\quad$ restore N to $\mathcal{O}$;
 $\quad\quad\quad$ **end**
 $\quad\quad$ **else if** $z_k < z^*$ **then**
 $\quad\quad\quad$ $z^* \leftarrow z_k$; $\boldsymbol{x}^* \leftarrow \boldsymbol{x}^k$; $\boldsymbol{w}^* \leftarrow \boldsymbol{w}^k$;
 $\quad\quad\quad$ $\boldsymbol{y}^* \leftarrow \boldsymbol{y}^k$, where $\boldsymbol{y}^k$ is a feasible solution of $\mathrm{SP}(\boldsymbol{x}^k)$
 $\quad\quad$ **end**
 $\quad$ **end**
 $\quad$ **else if** $z_k \leq z^*$ **then**
 $\quad\quad$ branch at N to create nodes N_1, N_2 and add them to $\mathcal{O}$;
 $\quad\quad$ select a fractional x_i^k;
 $\quad\quad$ add $x_i \leq \lfloor x_i^k \rfloor$, $x_i \geq \lceil x_i^k \rceil$ to the LP relaxation at N_1, N_2 respectively
 $\quad$ **end**
 end
end
if $z^* < \infty$ **then** $(\boldsymbol{x}^*, \boldsymbol{y}^*, \boldsymbol{w}^*)$ is an optimal solution;
else (4.1) is infeasible;

Algorithm 1: Generic branch-and-check algorithm when the master problem is an MILP with integer variables $\boldsymbol{x}$ and continuous variables $\boldsymbol{w}$

Figure 4.1 shows a branch-and-check tree for the assignment and scheduling example of Sect. 1.4. The tree is traversed in a depth-first fashion, taking the left branch first. The LP relaxation of the master problem is obtained by replacing $x_{ij} \in \{0, 1\}$ with $0 \leq x_{ij} \leq 1$ for all i, j. We also impose the bound $M \geq 0$ on the makespan to avoid an unbounded solution of the master problem in the first iteration. In the figure, M_k is the optimal value z_k of the current

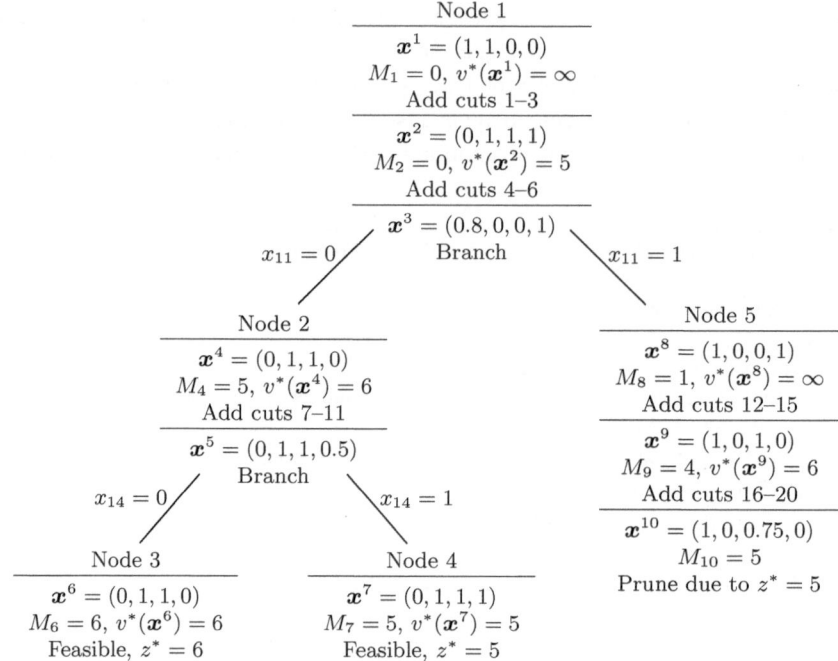

Fig. 4.1 Example of branch and check. Cuts are listed in Table 4.1

master problem, and $\boldsymbol{x}^k$ refers to the tuple $(x_{11}^k, x_{12}^k, x_{13}^k, x_{14}^k)$. The subproblem $SP(\boldsymbol{x}^k)$ is solved when $\boldsymbol{x}^k$ is integral, and cuts generated when $M_k < v^*(\boldsymbol{x}^k)$. The Benders cuts are referenced by number and are listed in Table 4.1. In this example, the procedure revisits a node immediately after cuts are added. This occurs twice at node 1, for example. There is no branching at node 5 because the current master problem value $M_{10} = 5$ is no better than the value $z^* = 5$ of the incumbent solution (the best solution so far). Since there are no open nodes at this point, the procedure terminates by identifying the incumbent solution $\boldsymbol{x}^* = (0, 1, 1, 1)$ and $\boldsymbol{s}^* = (3, 3, 0, 4)$ as optimal.

Branch and check might appear to be a special case of branch and cut when applied to an MILP master problem, but it is not. It uses Benders cuts that differ from traditional cuts in two ways. First, they are not are valid for the master problem, but are derived from the subproblem. Second, the variables they contain all have integer values in the current solution of the LP relaxation, whereas a traditional cut contains at least one variable with a fractional value, since otherwise the cut would not have been generated.

Branching methods typically use a primal heuristic to generate feasible solutions at the root node of the search tree, and perhaps at other nodes. These feasible solutions can be used to obtain additional Benders cuts, sometimes to great advantage. Another practical consideration is that branch and check requires one to modify the code that solves the

Table 4.1 Benders cuts used in the branch-and-check tree of Fig. 4.1

No	Shop	Benders cut	No	Shop	Benders cut
1	1	$x_{11} + x_{12} \leq 1$	11	2	$M \geq 6 - 3(1 - x_{21}) - 2(1 - x_{24})$
2	2	$M \geq 4x_{24}$	12	1	$x_{11} + x_{14} \leq 1$
3	2	$M \geq 4 - 2(1 - x_{23}) - 3(1 - x_{24})$	13	2	$M \geq 4x_{22}$
4	1	$M \geq 5(x_{12} + x_{14} - 1)$	14	2	$M \geq 3 - 3(1 - x_{22}) - 2(1 - x_{23})$
5	1	$M \geq 5 - (1 - x_{12}) - 3(1 - x_{13}) - (1 - x_{14})$	15	2	$M \geq 4 - 4(1 - x_{22}) - 3(1 - x_{23})$
6	2	$M \geq 5x_{21}$	16	1	$M \geq 6(x_{11} + x_{13} - 1)$
7	1	$M \geq 4(x_{12} + x_{13} - 1)$	17	1	$M \geq 5 - 3(1 - x_{11}) - 3(1 - x_{13})$
8	1	$M \geq 4 - 3(1 - x_{12}) - 3(1 - x_{13})$	18	1	$M \geq 6 - 4(1 - x_{11}) - 4(1 - x_{13})$
9	2	$M \geq 6(x_{21} + x_{24} - 1)$	19	2	$M \geq 5(x_{22} + x_{24} - 1)$
10	2	$M \geq 5 - 2(1 - x_{21}) - (1 - x_{24})$	20	2	$M \geq 3(1 - x_{22}) - 3(1 - x_{24})$

master problem, or at least to use a code that allows callbacks to insert new constraints at each node. This contrasts with standard LBBD, which can use an off-the-shelf MILP solver without callbacks. Branch and check can therefore take longer to implement.

4.3 Enumerative LBBD

Benders optimality cuts impose a bound on the optimal value, while feasibility cuts explicitly exclude solutions. A feasibility cut is sometimes more convenient even when the subproblem is an optimization problem. For example, combinatorial Benders cuts (Sect. 4.4 below) are always feasibility cuts. An *enumerative* method addresses this issue. It makes a feasible subproblem infeasible by requiring that the subproblem value be better that its current optimal value. It then generates a feasibility cut for the resulting infeasible subproblem. Cuts are generated until the master problem becomes infeasible, indicating that all feasible solutions have been (implicitly or explicitly) enumerated.

Enumerative LBBD is defined for problems in which the objective value depends only on subproblem variables y, although we will see that this restriction can be circumvented by a minor reformulation of the problem. We therefore address problems of the form

$$\min_{x, w, y} \left\{ f(y) \mid \mathcal{C}(x, y), \; \mathcal{C}'(x, w), \; x \in \mathcal{D}_x, \; y \in \mathcal{D}_y, \; w \in \mathcal{D}_w \right\} \qquad (4.3)$$

The master problem in iteration k is the feasibility problem

$$\{(x, w) \mid C'(x, w);\ F_{x\ell}(x),\ \ell = 1, \ldots, k - 1;\ x \in \mathcal{D}_x,\ w \in \mathcal{D}_w\} \tag{4.4}$$

where $F_{x\ell}(x)$ for $\ell = 1, \ldots, k - 1$ are feasibility cuts from previous iterations. Given a solution x^k of the master problem, the subproblem $\overline{\text{SP}}(x^k)$ has optimal value $v^*(x^k)$ and can be stated

$$\min_y \{f(y) \mid C(x^k, y),\ f(y) < v^*,\ y \in \mathcal{D}_y\} \tag{4.5}$$

where v^* (initially ∞) is the best subproblem value found so far. If $\overline{\text{SP}}(x^k)$ is infeasible, a feasibility cut $F_{x^k}(x)$ is generated. If $\overline{\text{SP}}(x^k)$ is feasible, v^* is reduced to $v^*(x^k)$, and a feasibility cut $F_{x^k}(x)$ is generated for the infeasible subproblem that results. The procedure continues until the master problem is infeasible. At this point, the original problem is infeasible if a feasible subproblem has not been found ($v^* = \infty$). Otherwise, the last feasible solution found is optimal. A formal statement of the procedure appears as Algorithm 2.

$k \leftarrow 0, v^* \leftarrow \infty$;
while *the master problem* (4.4) *is feasible* **do**
 $k \leftarrow k + 1$;
 let (x^k, w^k) be a solution of (4.4);
 solve $\overline{\text{SP}}(x^k)$ and let $v_k^*(x^k)$ be its optimal value;
 if $\overline{\text{SP}}(x^k)$ *is infeasible* **then** generate feasibility cut $F_{x^k}(x)$;
 else
  let y^* be an optimal solution of $\overline{\text{SP}}(x^k)$;
  $v^* \leftarrow v^*(x^k)$; generate feasibility cut $F_{x^k}(x)$ for $\overline{\text{SP}}(x^k)$;
  $x^* \leftarrow x^k$; $w^* \leftarrow w^k$
 end
end
if $v^* < \infty$ **then** (x^*, y^*, w^*) solves (4.3);
else (4.3) is infeasible;

Algorithm 2: Enumerative LBBD procedure

One way to obtain a feasibility cut for a feasible subproblem is to derive an optimality cut $z \geq B_{x^k}(x)$ for $\overline{\text{SP}}(x^k)$ before v^* is updated and convert it to a feasibility cut. One need only write a feasiblity cut that excludes values of x for which $B_{x^k}(x) = B_{x^k}(x^k)$. For example, a nogood optimality cut

$$z \geq v^*(\bar{x}) - \left(v^*(\bar{x}) - \underline{v}\right)\left(\sum_{j \in J_1(\bar{x})}(1 - x_j) + \sum_{j \in J_0(\bar{x})} x_j\right)$$

can be converted to the feasibility cut

$$\sum_{j \in J_1(\bar{x})} (1 - x_j) + \sum_{j \in J_0(\bar{x})} x_j \geq 1$$

However, conversion to a feasibility cut generally loses information when $z \geq B_{x^k}(x)$ is an analytical or multivalent cut, because such a cut provides useful bounds for values of x not excluded by a feasibility cut.

An enumerative method may partially sacrifice the advantage of subproblem decoupling, in two ways. One is that the constraint $f(y) < v^*$ must be enforced for the subproblem as a whole, and this may require recoupling of the subproblem components. Another is that an optimality cut $z \geq B_{x^k}(x)$ can be converted to a feasibility cut only if it the optimality cut is valid for the entire subproblem. An optimality cut $z_i \geq B^i_{x^k}(x)$ that is valid for a subproblem component i (using the notation of Sect. 2.6) need not be a valid Benders cut for the subproblem as a whole, because it may violate condition (B2). On the other hand, if a subproblem component is infeasible, then the feasibility cut it generates is valid for the entire subproblem, even though it is based only on one component of the subproblem. Thus the advantage of decoupling is partially retained.

These ideas can be illustrated by the minimum makespan problem of Sect. 1.4, in which $x_{ij} = 1$ when job j is assigned to shop i. In that problem, the makespan depends only on the start times s_j of jobs j, variables that occur only in the scheduling subproblem. The problem is therefore a candidate for enumerative LBBD. The procedure, which is summarized in Table 4.2, goes as follows. We found in Sect. 1.4 that when the master problem initially assigns jobs 1 and 2 to shop 1 (iteration 1), that shop has no feasible schedule, and shop 2 has a minimum makespan of 4. The feasiblity cut $x_{11} + x_{12} \leq 1$ generated by shop 1 is valid for the subproblem as a whole, even though it is based only on shop 1. However, the optimality cut $M \geq 4x_4$ generated by shop 2 is not valid for the entire subproblem, for which

Table 4.2 Enumerative Benders iterations for the example problem of Sect. 1.4

Iteration	Master problem solution $(x_{11}, \ldots, x_{14})$	Sub-problem optimal value[†]	Subproblem solution[†] $(s_1, \ldots, s_4)$	Updated v^*
1	$(1, 1, 0, 0)$	∞	None	∞
2	$(0, 1, 0, 1)$	5	$(3, 3, 0, 4)$	5
3	$(1, 0, 0, 0)$	∞	None	5
4	$(0, 0, 1, 1)$	∞	None	5
5	$(0, 1, 1, 0)$	∞	None	5
6	None			

[†]before updating v^*

the minimum makespan is ∞, not 4. We therefore cannot update v^* from ∞ to 4, and the optimality cut cannot be converted to a feasibility cut $x_4 \leq 0$.

At this point, the master problem contains the single cut $x_{11} + x_{12} \leq 1$ (iteration 2). The feasible solution $(x_{11}, \ldots, x_{14}) = (0, 1, 0, 1)$ results in the scheduling subproblem of Fig. 1.2, in which shops 1 and 2 have a minimum makespan of 5 and 4, respectively. Since the entire subproblem has a minimum makespan of 5, a feasibility cut can be generated for shop 1 but not shop 2. We update v^* to 5 and impose the feasiblity cut $x_{12} + x_{14} \leq 1$. Since we are minimizing makespan, the constraint $f(s) < 5$ is easily enforced in future subproblems by enforcing it in individual shops. The solution $(1, 0, 0, 0)$ is feasible in the new master problem (iteration 3), which leads to the scheduling problems of Fig. 1.3, with the added constraint that the makespan in both shops must be less than 5. Neither shop has a feasible schedule, and the feasibility cuts generated by each is a valid Benders cut for the entire subproblem: $x_{11} \leq 0$ and $x_{12} + x_{14} \geq 1$, respectively. Two more master problem solutions (iterations 4 and 5) lead to infeasiblity of the master problem in iteration 6 as shown in Table 4.2. The optimal solution is the last (and only) feasible solution obtained (iteration 2).

An enumerative procedure cannot be directly applied when the objective function depends on master problem variables as well as subproblem variables. However, this restriction can be circumvented in principle by duplicating the master problem variables in the subproblem. Thus a problem of the form

$$\min_{x,y,w} \left\{ f(x, y) \mid C(x, y), \ C'(x, w), \ x \in \mathcal{D}_x, \ y \in \mathcal{D}_y, \ w \in \mathcal{D}_w \right\}$$

is reformulated

$$\min_{x,y,w,\xi} \left\{ f(\xi, y) \mid C(x, y), \ C'(x, w), \ x = \xi, \ x \in \mathcal{D}_x, \ y \in \mathcal{D}_y, \ w \in \mathcal{D}_w \right\}$$

where ξ and y are viewed as subproblem variables. The subproblem $\overline{SP}(x^k)$ in iteration k is

$$\min_{y,\xi} \left\{ f(\xi, y) \mid C(x^k, y), \ \xi = x^k, \ f(\xi, y) < v^*, \ y \in \mathcal{D}_y \right\}$$

The practical effect of this is that the new constraints $\xi = x^k$ must be taken into consideration when valid Benders cuts are obtained. This may complicate the structure of the subproblem and make the generation of effective cuts more difficult.

4.4   Combinatorial Benders Cuts for MILP

Combinatorial Benders cuts are cuts used in a specialized branch-and-check method for mixed integer/linear programming (MILP) models [60]. It places all integer variables in the master problem, and all continuous variables in a linear programming (LP) subproblem.

It assumes that the subproblem is linked to the master problem by conditional constraints that activate subproblem constraints when corresponding 0–1 variables are equal to 1. The conditional constraints often take the form of "big-M" constraints in MILP models. It is also assumed that the objective function depends only on the integer variables, making the subproblem a feasibility problem, although this restriction can be partially circumvented by an enumerative variant of the method.

While most LBBD cuts are combinatorial in some sense, we follow the convention of referring to cuts as "combinatorial" only when they occur in the specialized method described in [60]. This particular specialization deserves attention for at least two reasons. It gets rid of troublesome big-M constraints, which are very common but whose weak LP relaxation makes MILP models harder to solve. At the same time, it allows one to design effective nogood cuts that are based on irreducible infeasible subsystems of the linear subproblem.

Problems suitable for combinatorial Benders cuts have the form

$$\min_{x,x',y} \left\{ cx + c'x' \; \middle| \; \begin{array}{l} a^r y \geq \beta_r - M_r(1 - x_{j(r)}), \; x_{j(r)} \in \{0, 1\}, \; \text{all } r \in R \\ Ax' \geq b, \; By \geq h, \; x' \in \mathbb{Z}^p, \; y \in \mathbb{R}^m \end{array} \right\}$$

where x is a tuple of 0–1 variables and x' a tuple of general integer variables. The constraint $a^r y \geq \beta_r - M_r(1 - x_{j(r)})$ is a big-M constraint that represents the conditional constraint $(x_{j(r)} = 1) \rightarrow (a^r y \geq \beta_r)$. The coefficients $M_r > 0$ are large enough so that $a^r y \geq \beta_r - M_r$ is satisfied by any y. A 0–1 variable can occur in two (or more) conditional constraints indexed by r and r' respectively, in which case we have $j(r) = j(r')$.

For a given solution $(\bar{x}, \bar{x}')$ of the master problem, the subproblem is the LP feasibility problem

$$\{y \in \mathbb{R}^m \mid a^r y \geq \beta_r, \; \text{all } r \in R(\bar{x}); \; By \geq h\} \tag{4.6}$$

where $R(\bar{x}) = \{r \mid \bar{x}_{j(r)} = 1\}$ is the index set of all constraints $a^r y \geq \beta_r$ enforced by $\bar{x}$. Since unenforced constraints are simply omitted from the subproblem, there is no longer a need for big-M's. The solution procedure is the same as for branch and check generally (Algorithm 1), except that the Benders cuts are always feasibility cuts because the subproblem is a feasibility problem.

When the subproblem is infeasible, we have a simple feasibility cut

$$\sum_{j \in J(\bar{x})} (1 - x_j) \geq 1$$

where $J(\bar{x}) = \{j(r) \mid r \in R(\bar{x})\}$. We can strengthen the cut by identifying an irreducible infeasible subset of constraints. We know that removing any constraint $a^r y \geq \beta_r$ from the irreducible subset relieves the infeasiblity. Thus, if $R'(\bar{x}) \subseteq R(\bar{x})$ indexes the constraints $a^r y \geq \beta_r$ that belong to the irreducible subset, we can write a strengthened nogood cut

$$\sum_{j \in J'(\bar{x})} (1 - x_j) \geq 1$$

where $J'(\bar{x}) = \{j(r) \mid r \in R'(\bar{x})\}$ is the index set of x_js that appear in the irreducible infeasible subset. A cut obtained in this fashion is a combinatorial Benders cut.

One way to identify an irreducible infeasible subset of constraints is to take advantage of information in the LP dual solution. Due to the Farkas lemma, any dual solution of an infeasible LP problem specifies a linear combination of constraints that proves infeasibility (Sect. 2.4). Only the constraints with positive multipliers in the dual solution serve as premises of the proof. In fact, it is easy to show that if the dual solution is an extreme ray, these premises are all essential to the proof, and the corresponding constraints comprise an irreducible infeasible set [54, 104]. Irreducible infeasible subsets for LP problems are further analyzed in [51–53, 172].

The LP dual of the subproblem (4.6) is

$$\max_{u,v} \left\{ \sum_{r \in R(\bar{x})} u_r \beta_r + vh \mid u_r a^r + vB = 0, \text{ all } r \in R(\bar{x}); \; u, v \geq 0 \right\} \qquad (4.7)$$

where dual variable u_r corresponds to $a^r y \geq \beta_r$ and the row vector v of dual variables corresponds to $By \geq h$. The dual problem is always feasible because $u = v = 0$ is a feasible solution, and it is therefore unbounded when the subproblem is infeasible. In practice, a dual solution (u, v) representing an extreme ray can be obtained by replacing the objective function with a normalizing constraint

$$\sum_{r \in R(\bar{x})} u_r \beta_r + vh = 1$$

and adding a heuristic objective function that tends to reduce the number of positive u_is and thereby produce a stronger nogood cut. This results in the revised dual problem

$$\min_{u,v} \left\{ \sum_{r \in R(\bar{x})} \gamma_r u_r \; \middle| \; \begin{array}{l} \sum_{r \in R(\bar{x})} u_r \beta_r + vh = 1; \; u, v \geq 0 \\ u_r a^r + vB = 0, \text{ all } r \in R(\bar{x}) \end{array} \right\}$$

Initially each $\gamma_r = 1$, but one may be able to obtain additional cuts by setting $\gamma_r = 0$ for some r.

When the objective function of the MILP problem depends on the continuous variables y rather than integer variables, the subproblem is no longer a feasibility problem. However, one can convert it to a feasibility problem for purposes of cut generation by using an enumerative method as described in the previous section. Now the problem is

$$\min_{x,x',y} \left\{ dy \; \middle| \; \begin{array}{l} a^r y \geq \beta_r - M_r(1 - x_j), \; x_{j(r)} \in \{0, 1\}, \text{ all } r \in R \\ Ax' \geq b, \; By \geq h, \; x' \in \mathbb{Z}^p, \; y \in \mathbb{R}^m \end{array} \right\}$$

and the subproblem (4.5) becomes

$$\min_y \left\{ \boldsymbol{d} \boldsymbol{y} \mid \boldsymbol{a}^r \boldsymbol{y} \geq \beta_r, \text{ all } r \in R(\bar{\boldsymbol{x}}); \ \boldsymbol{B} \boldsymbol{y} \geq \boldsymbol{h}; \ \boldsymbol{d} \boldsymbol{y} \leq v^* - \epsilon; \ \boldsymbol{y} \in \mathbb{R}^m \right\} \qquad (4.8)$$

where v^* is the best subproblem value found so far. If the subproblem is infeasible, a nogood feasibility cut is generated by identifying a minimal infeasible subset of the constraints, which now include the bound $\boldsymbol{d} \boldsymbol{y} \leq v^* - \epsilon$. If the subproblem is feasible, let $v^*(\bar{\boldsymbol{x}})$ be its optimal value, and $\bar{\boldsymbol{y}}$ its optimal solution. If $v^*(\bar{\boldsymbol{x}}) < v^*$, record $(\bar{\boldsymbol{x}}, \bar{\boldsymbol{x}}', \bar{\boldsymbol{y}})$ as the incumbent solution in the branch-and-check procedure, and update v^* by setting $v^* = v^*(\bar{\boldsymbol{x}})$. The latter action makes the subproblem infeasible, and a nogood feasiblity cut can be generated, again by identifying a minimal infeasible subsystem. The branch-and-check procedure continues until the enumeration tree is exhausted, at which point the incumbent solution is optimal.

If the objective function depends on the 0–1 variables as well as the continuous variables, we can duplicate the 0–1 variables in the subproblem, much as described at the end of the previous section. Thus if the objective function is $\boldsymbol{c} \boldsymbol{x} + \boldsymbol{d} \boldsymbol{y}$, variables x_j are duplicated in the subproblem as ξ_j. The constraints $\boldsymbol{x} = \boldsymbol{\xi}$ can be written as additional conditional constraints $x_j - \xi_j \geq 0$ and $-x_j + \xi_j \geq 0$. The subproblem (4.8) becomes

$$\min_{y, \xi} \left\{ \boldsymbol{c} \boldsymbol{\xi} + \boldsymbol{d} \boldsymbol{y} \left| \begin{array}{l} \boldsymbol{a}^r \boldsymbol{y} \geq \beta_r, \text{ all } r \in R(\bar{\boldsymbol{x}}); \ \boldsymbol{B} \boldsymbol{y} \geq \boldsymbol{h} \\ \xi_j \geq 1, \ -\xi_j \geq -1, \text{ all } j \in J(\bar{\boldsymbol{x}}) \\ -\xi_j \geq 0, \ \xi_j \geq 0, \text{ all } j \notin J(\bar{\boldsymbol{x}}) \\ \boldsymbol{c} \boldsymbol{\xi} + \boldsymbol{d} \boldsymbol{y} \leq v^* - \epsilon, \ \boldsymbol{\xi} \in \mathbb{R}^n, \ \boldsymbol{y} \in \mathbb{R}^m \end{array} \right. \right\}$$

The new constraints $\xi_j \geq 1$, $-\xi_j \geq -1$, $\xi_j \geq 0$, and $-\xi_j \geq 0$, must be taken into account when identifying a minimal infeasible subsystem.

When the objective function depends on general integer variables as well as 0–1 variables, it is less obvious how to develop combinatorial Benders cuts for the problem.

4.5   Stochastic and Robust Optimization

Benders decomposition provides a natural approach to two-stage stochastic and robust optimization problems, because the second stage decouples into smaller problems. The objective in such problems is to minimize expected or worst-case cost when some of the problem parameters are uncertain and take different values in different scenarios. Decisions in the first stage determine which scenarios are possible in the second stage, where an optimal solution is computed for each of the scenarios. The scenario-specific problems can be solved independently, because all linking variables are confined to the first stage problem. This presents an ideal opportunity for Benders decomposition, because the second stage problem can be viewed as a Benders subproblem, where decoupling confers a substantial advantage when there are hundreds or thousands of scenarios. LBBD comes into play when the second stage involves integer variables ("integer recourse") or other discrete variables.

Integer variables in the second stage are historically accommodated by the *integer L-shaped method* [8, 159], a variation of classical Benders decomposition that uses classical cuts derived from the LP relaxation of the subproblem, as well as "integer cuts" to ensure convergence. However, the integer cuts are simple nogood cuts that are quite weak. In addition, the LP relaxation required for classical cuts is based on an MILP formulation of the subproblem, which often has a weak LP relaxation and consequently yields weak classical cuts. Solution of the LP relaxation can also add significant computational overhead. LBBD replaces the simple nogood cuts with stronger logic-based cuts and dispenses with the LP subproblem relaxation, often resulting in substantial computational speedups [75].

Two-stage stochastic optimization strives to minimize expected cost over all scenarios, assuming that cost is minimized in each scenario [26, 206, 229]. We assume for ease of exposition that the set Ω of scenarios is finite. If we suppose that scenario ω has probability π_ω, the problem has the form

$$\min_{x} \left\{ g(x) + \sum_{\omega \in \Omega} \pi_\omega h_\omega^*(x) \;\middle|\; C'(x), \; x \in \mathcal{D}_x \right\} \qquad (4.9)$$

The second stage problem is to calculate

$$h_\omega^*(x) = \min_{y} \left\{ h_\omega(x, y) \;\middle|\; C_\omega(x, y), \; y \in \mathcal{D}_y \right\}$$

If we create a copy y^ω of y for each scenario ω, problem (4.9) can be written

$$\min_{x, y} \left\{ g(x) + \sum_{\omega \in \Omega} \pi_\omega h_\omega(x, y^\omega) \;\middle|\; \begin{array}{l} C_\omega(x, y), \; y^\omega \in \mathcal{D}_y, \text{ all } \omega \in \Omega \\ C'(x), \; x \in \mathcal{D}_x \end{array} \right\}$$

This problem has the form of an optimization problem (2.13) with a subproblem that decouples, where

$$f(x, y) = g(x) + \sum_{\omega \in \Omega} \pi_\omega h_\omega(x, y^\omega), \quad y^i = y^\omega, \quad \text{and} \quad C'(x, w) = C'(x)$$

Problem (4.9) can therefore be solved by LBBD as described in Sect. 2.6.

In a two-stage robust optimization problem, Ω is an uncertainty set that consists of reasonably possible scenarios [24, 25]. The objective is to minimize the worst-case optimal cost over the scenarios:

$$\min_{x} \left\{ g(x) + \max_{\omega \in \Omega} \left\{ h_\omega^*(x) \right\} \;\middle|\; C'(x), \; x \in \mathcal{D}_x \right\}$$

The problem can be written

$$\min_{x, y} \left\{ g(x) + \max_{\omega \in \Omega} \left\{ h_\omega(x, y^\omega) \right\} \;\middle|\; \begin{array}{l} C_\omega(x, y), \; y^\omega \in \mathcal{D}_y, \text{ all } \omega \in \Omega \\ C'(x), \; x \in \mathcal{D}_x \end{array} \right\}$$

which again has the form (2.13) and can be solved by LBBD.

4.6 Subproblem Relaxation in the Master Problem

In standard Benders decomposition, the master problem receives information from the sub-problem only in the form of Benders cuts. However, there is nothing in the concept of decomposition that prevents a richer transfer of information. Recall that cuts generated during a Benders algorithm partially define the projection of the epigraph onto the master problem variables. Since a complete description of the projection can require a vast number of Benders cuts, relatively few of the more relevant cuts are generated in response to master problem solutions. Yet, if there is a reasonably small family of valid constraints on the projection that can be identified in advance, there is no reason they cannot be included in the master problem at the outset.

We refer to this strategy as *including a subproblem relaxation in the master problem*. Experience teaches that it is often an essential component of a successful LBBD application. One variation of the strategy is to duplicate some of the subproblem constraints in the master problem, perhaps while relaxing the integrality constraint on subproblem variables. Another is to formulate some valid constraints that are implied by the subproblem but contain only master problem variables. We consider each of these in turn.

4.6.1 Duplicating Subproblem Constraints

A subproblem relaxation is frequently included in the master problem simply by duplicating some subproblem constraints in the master problem. These constraints bring subproblem variables along with them, seemingly in violation of the principle that Benders decomposition partitions the variables. However, this principle is formally accommodated by permitting auxiliary variables w_j in the master problem to represent corresponding subproblem variables y_j in the imported constraints.

It is partly to enable this maneuver that auxiliary variables w appear in the general problem statement introduced in Sect. 2.3:

$$\min_{x,y,w} \left\{ f(x, y) \mid C(x, y), \ C'(x, w), \ x \in \mathcal{D}_x, \ y \in \mathcal{D}_y, \ w \in \mathcal{D}_w \right\}$$

The constraint set $C'(x, w)$ can include subproblem constraints, which therefore appear in the master problem:

$$\min_{z,x,w} \left\{ z \mid C'(x, w); \ z \geq B_{x^\ell}(x), \ \ell = 1, \ldots, k-1; \ x \in \mathcal{D}_x, \ w \in \mathcal{D}_w \right\}$$

While constraints in $C'(x, w)$ can contain variables in x or w or both, subproblem constraints in $C'(x, w)$ would, of course, contain only variables in w. Another point to remember is that the solution values of variables in w are not passed to the subproblem. Thus, the subproblem is solved for its variables y without knowledge of the values obtained for corresponding variables in w.

Obviously, including too much of the subproblem in the master problem defeats the purpose of decomposition. It results in a master problem that closely resembles the original problem and is therefore hard to solve. In fact, since the subproblem is typically a combinatorial problem in LBBD applications, duplication of even a very few key subproblem constraints can make the master problem intractable. A popular solution to this dilemma is to relax any integrality constraints on the variables in w that represent subproblem variables. For example, variables $y_j \in \{0, 1\}$ in the subproblem could be represented by auxiliary variables $w_j \in [0, 1]$, and the domain $\mathcal{D}_w$ defined accordingly. In some applications, *all* of the subproblem constraints can be duplicated in the master problem with good computational results.

4.6.2   Subproblem Relaxation Without Subproblem Variables

A second method of including a subproblem relaxation in the master problem is to identify valid inequalities for the original problem (before decomposition) and select a family of inequalities that contain only master problem variables. This provides information about the subproblem without introducing subproblem variables into the master problem. Decades of research have identified valid inequalities for a wide range of combinatorial optimization problems, and one frequent strategy is to draw on this research.

However, valid inequalities enable communication with the subproblem only if they are based at least partly on subproblem constraints, and such inequalities often contain subproblem variables. In such cases, it can be useful to derive a relaxation *de novo* that is based on the subproblem but contains only master problem variables. This popular strategy is most often implemented by deriving a relaxation from time window constraints in a scheduling subproblem. We illustrate how this can be accomplished.

Using the notation of Sect. 3.4, suppose there are n jobs to be scheduled, where each job j has processing time p_j and time window $[r_j, d_j]$. The binary master problem variable $x_j = 1$ when job j is selected to be scheduled. For the sake of generality, we suppose the subproblem is a cumulative scheduling problem in which the maximum permissible rate of resource consumption is C, and each job j consumes resources at the rate c_j. In a disjunctive scheduling problem, $C = 1$ and each $c_j = 1$. If we let the *energy* of job j be $p_j c_j$, the total energy of jobs scheduled between times t_1 and t_2 can be at most $C(t_2 - t_1)$. This yields the valid relaxation

$$\sum_{j \in J} p_j c_j x_j \leq C(L_J - E_J)$$

for any set J of jobs, where $E_J = \min_{j \in J}\{r_j\}$ is the earliest release time and $L_J = \max_{j \in J}\{d_j\}$ the latest deadline of the jobs in J. This relaxation can be added to the master problem when J is the set of all jobs, or perhaps a subset of jobs whose time windows tend to overlap (to obtain a tighter relaxation).

In a minimum makespan problem, we can bound the minimum makespan by observing how much time is required to accommodate the energy of selected jobs. Thus if z represents the makespan, we have the valid relaxation

$$z \geq E_J + \frac{1}{C} \sum_{j \in J} p_j c_j x_j$$

where $(1/C) \sum_j p_j c_j x_j$ is a lower bound on the time required to run the jobs in J.

If we are minimizing the number of late jobs, we can observe that if this lower bound exceeds the available time $L_J - E_J$, one or more jobs will necessarily be late. The number of late jobs is at least the excess time divided by the maximum processing time over all the jobs. Thus if z represents the number of late jobs, we have the valid bound

$$z \geq \frac{\frac{1}{C} \sum_{j \in J} p_j c_j x_j - (L_J - E_J)}{\max_{j \in J}\{p_j\}}$$

If we are minimizing total tardiness, at least two valid relaxations are available, neither of which dominates the other [123]. Let $J_k = \{j \in J \mid d_j \leq d_k\}$ be the set of jobs in J with due date no later than job k's due date. Then if only jobs from J_k are scheduled, the job with the latest completion time can finish no earlier than $E_J + \sum_{i \in J_k} p_j c_j x_j$. This means that its tardiness is no less than

$$\left(E_J + \sum_{j \in J_k} p_j c_j x_j - d_k\right)^+$$

Total tardiness is therefore no less than this amount. Thus, if z represents total tardiness, we can write a relaxation consisting of the bounds

$$z \geq \left(E_J + \sum_{j \in J_k} p_j c_j x_j - d_k\right)^+, \quad k = 1, \ldots, n$$

A second relaxation can be obtained if we index the jobs so that $d_1 \leq \cdots \leq d_n$. We also let $\pi_1, \ldots, \pi_n$ be a permutation of $1, \ldots, n$ that orders the jobs by nondecreasing energy; that is, $p_{\pi_1} c_{\pi_1} \leq \cdots \leq p_{\pi_n} c_{\pi_n}$. Then if we introduce continuous variables $z_1, \ldots, z_n$, we have a relaxation consisting of

$$z \geq \sum_{k=1}^{n} z_k x_k$$

$$z_k \geq E_J + \frac{1}{C} \sum_{j=1}^{k} p_{\pi_j} c_{\pi_j} x_{\pi_j} - d_k, \ z_k \geq 0, \quad k = 1, \ldots, n$$

The terms $z_k x_k$ are nonlinear, but we can linearize the relaxation by writing

$$z \geq \sum_{k=1}^{n} z_k$$

$$z_k \geq E_J + \frac{1}{C} \sum_{j=1}^{k} p_{\pi_j} c_{\pi_j} x_{\pi_j} - d_k - (1 - x_k) M_k, \ z_k \geq 0, \quad k = 1, \ldots, n$$

where

$$M_k = E_J + \frac{1}{C} \sum_{j=1}^{k} p_{\pi_j} c_{\pi_j} - d_k, \quad k = 1, \ldots, n$$

The big-M term M_k can be negative, but the lower bound on z_k in the relaxation is never positive when $x_k = 0$.

Validity proofs of the two relaxations appear in [123]. Proofs and examples are given in Sect. 7.13 of [124]. A number of other, problem-specific relaxations can be found in the application articles summarized in Chap. 5.

4.7   Multilevel LBBD

It is occasionally helpful in practice to apply three-level LBBD, rather than the standard two-level method, as for example in [21, 22, 58, 218, 255]. One can experiment with this option when both the master problem and subproblem remain intractable for any reasonable partition of the variable set, to determine whether a three-way partition leads to solubility.

Three-level LBBD is implemented simply by solving the subproblem with a second LBBD procedure. For this purpose, the subproblem variables are partitioned for inclusion in a secondary master problem and subproblem. Cuts from the secondary subproblem are added to the secondary master problem until the original subproblem is solved, whereupon cuts are generated for the original master problem. If desired, either the primary or secondary procedure can be branch and check.

One can, in principle, solve even the secondary subproblem by LBBD, and so on recursively, although there appear to be no published reports in which this is found useful. A theoretical study of multilevel Benders decomposition is provided in [28].

4.8   Dynamic Variable Partitioning

In standard Benders algorithms, there is a fixed partition of the variable set into master problem and subproblem variables. Yet, the partition can equally well change from one iteration to the next, so long as the Benders cuts satisfy properties (B1) and (B2) stated in Sect. 2.3.

The argument for this goes as follows. The Benders cuts are valid due to property (B1). If the usual termination condition is used, an LBBD algorithm that terminates must deliver an optimal solution (or prove infeasibility), due to the validity of the cuts. Finite convergence is likewise guaranteed, due to virtually the same argument used in the proof Theorem 2.1: a given assignment x^k of values to a given set of master problem variables x cannot repeat before the final iteration. This is because x^k can be optimal in the master problem a second time (say, iteration $k' > k$) only if the optimal value $z_{k'}$ of the master is $v^*(x^k)$, whereupon the algorithm terminates. We must have $z_{k'} = v^*(x^k)$ because the cut generated by x^k imposes the lower bound $v^*(x^k)$ on the master problem value when $x = x^k$, due to property (B2). Since there are a bounded number of ways to select master problem variables x, and a bounded number of values x^k they can be assigned in any iteration k (due to finite domains), the algorithm must terminate after a bounded number of iterations.

Dynamic variable partitioning is most naturally used in a branch-and-check context. For example, one might invoke the subproblem when only a proper subset of the integer variables have integer values at a node of the search tree, leaving the other integer variables to appear (at least implicitly) in the subproblem.

Branch and check with dynamic variable partitioning is the most popular scheme used in state-of-the-art satisfiability (SAT) solvers, where conflict-directed clause learning is a key element [19]. The constraint set consists of logical clauses, such as $x_1 \vee \neg x_2 \vee x_3$. The master problem is solved by branching on the x_is in a Davis-Putnam-Loveland-Logemann (DPLL) procedure, or by some other search mechanism. At each branching node, the variables fixed so far become premises in a unit propagation algorithm (also known as unit resolution), an incomplete inference method. If unit propagation proves unsatisfiability of the clause set, the problem of proving unsatisfiability (given the fixed variables) can be regarded as a feasiblity subproblem for which unit propagation is a complete inference method. The unit propagation proof is represented as an implication graph from which conflict clauses are derived, much as Sect. 3.8 illustrates in a more general context. The conflict clauses are then treated as logic-based Benders cuts that satisfy property (B2) because unit propagation is complete for the subproblem. The variable partition is therefore dynamic, because it depends on which variables are fixed when unit propagation derives unsatisfiability.

4.9   Off-the-Shelf LBBD

It would undeniably be convenient if LBBD were available in off-the-shelf solvers, as is the
case for mixed integer/linear programming (MILP) and propositional satisfiability (SAT).
This would obviate the necessity of developing hand-crafted Benders cuts—provided, of
course, that the solver is capable of generating cuts that are equally adept at exploiting
problem structure.

One approach to implementing LBBD in a general-purpose solver is to maintain a library
of logic-based cuts that have been developed in advance, much as MILP solvers include a
library of cutting planes. Yet, the solver must somehow select cuts that are suitable for a
given problem. If the model is written using high-level global constraints, as in constraint
programming (CP), each constraint could be associated with cuts that would be activated
if that constraint appears in the subproblem. This strategy is pursued in the experimental
solver SIMPL [9, 262], which allows such high-level "metaconstraints" as fixed charge
network flows and cumulative scheduling to invoke suitable Benders cuts. This is practical,
however, only if the subproblem can be represented by known metaconstraints, and the
human modeler takes advantage of this. When low-level modeling is used, the solver must
recognize problem substructure automatically. This is done to some extent in MILP solvers
for the generation of cutting planes, but given current technology, it is a challenging task for
the much larger collection of logic-based Benders cuts.

An alternative strategy is to return to the root conception of logic-based cuts as derived
from the optimality proof obtained for the subproblem. The subproblem solver would be
expected to provide a trace of the optimality proof from which cuts could be constructed. The
idea is illustrated in Sect. 3.8, where an implication graph and nogood cuts are recovered from
the solution process, in much the way conflict clauses are derived in SAT solvers. Since the
subproblem will almost certainly be submitted to a general-purpose solver, the exploitation
of special structure is limited to what is already carried out in the solver. Nonetheless, an
off-the-shelf LBBD solver based on this approach could be useful as a method of first resort.
The situation is similar in MILP, where specialized cuts are typically designed only after
general-purpose solvers fail.

At this writing, LBBD is automated in the solver Nutmeg [154], which was built jointly by
two teams that independently developed precursors in [68, 155]. In one of the earlier efforts
[155], logic-based cuts are automatically generated in a branch-and-check procedure to solve
the capacitated traveling salesman problem with time windows. Cuts are obtained from a
CP solver roughly in the way described in Sect. 3.8, using a process that is quite general and
can be applied to any subproblem solved by CP. In [68], branch and check is incorporated
into the MiniZinc constraint modeling language. The reformulation mechanism of MiniZinc
selects constraints for an MILP master problem and a CP subproblem. The CP solver again
provides nogood cuts in somewhat the manner described in Sect. 3.8. Nutmeg is a stand-alone
branch-and-cut solver that, like MiniZinc, automatically partitions constraints between an
MILP master program and a CP subproblem. It inherits its cut generation method from its

two predecessors. Exploitation of special structure is otherwise limited; the subproblem, for example, is not decoupled even when this is possible. Nonetheless, an automatic LBBD solver of this kind can offer a substantial advantage relative to pure MILP or CP.

A attractive prospect for the future is the use of artificial intelligence to recognize problem structure that is suitable for various types of decomposition and identify appropriate specialized cuts on that basis. Machine learning is already being used to augment MILP solvers, and similar techniques may be adaptable to LBBD solvers.

Applications

5

5.1 Introduction

This final chapter surveys applications of LBBD and its variations as reported in the technical literature. The aim is to provide a repository of suggestions as to how LBBD can be adapted to a wide variety of problem domains. Brief descriptions are given for each application, often covering multiple relevant articles. The descriptions focus primarily on how the master problem and subproblem are defined, and the types of logic-based Benders cuts used. Readers should consult the original articles for full technical details, as only the basic ideas are provided here.

While an attempt is made to arrange the material by application domain, a given problem can frequently be classified in multiple ways. What seems to be the most helpful classification is used here. In addition, problems in very different application areas can have similar mathematical structure and may therefore yield to similar solution methods.

The following abbreviations are used throughout the chapter:

MP—master problem
SP—subproblem
CP—constraint programming
LP—linear programming
MILP—mixed integer/linear programming

The discussion of Benders cuts presupposes familiarity with the material in previous chapters, and cross-references to relevant sections are frequently provided. A glossary of logic-based cuts can be found in Table 3.1 of Chap. 3.

5.2   Transportation

5.2.1   Vehicle Routing

- **Single vehicle routing with capacity constraints and time windows** [155]. The problem is to route a capacitated vehicle from one customer to another so as to minimize total travel time subject to time windows. Branch and check (Sect. 4.2) is used to solve the two-index model of the problem. In the MP, 0–1 variables $x_{ij} = 1$ when the vehicle traverses the arc from customer i to customer j. The MP is an MILP model that contains only node degree constraints: the vehicle must enter and exit each node exactly once. The SP uses a CP solver with conflict analysis to check for time window violations. It deploys a general-purpose cut-generating method based on the solver's conflict analysis, similar to that described in Sect. 3.8. This, along with the automatic LBBD implementation in MiniZinc [68], led to the general-purpose LBBD solver Nutmeg [154] as recounted in Sect. 4.8.
- **Robust vehicle routing with time windows** [3]. Two models are proposed for adjustable robust optimization with uncertain travel times (Sect. 4.5). The MP assigns origin-destination paths to each vehicle. Each scenario corresponds to a vertex of a polyhedral uncertainty set. The SP solves a set of LPs to determine whether time windows are observed in every scenario. Strengthened feasibility cuts are used. The strengthening method is unclear, but see Sect. 3.5.2 for strong analytical cuts that could be used.
- **Travelling purchaser problem** [29]. The MP determines a set of markets and directed travel arcs that satisfy product purchase constraints with relaxed travel costs. The SP finds a tour that covers the selected markets and generates a cut in the form of a subtour elimination constraint for the complete graph induced by the vertices in the tour (slightly modified for a traveling purchaser). Both standard LBBD and branch-and-check (Sect. 4.2) methods are applied. This can be viewed as LBBD rather than lazy constraint generation because the MP omits variables used to enforce subtour elimination.
- **Set a price to visit customers, based on optimal routing** [1]. A company wishes to set a price for visiting a customer to install an appliance or perform some other service, based on the probability that a customer will be willing to pay the price. It is assumed that an optimal route will be used once the price is determined. The price a given customer is willing to pay is a random variable. The MP computes the expected revenue for a given price using linear chance constraints, decides on the price, and determines which customers to visit. Thus all stochastic elements are in the MP, and there is no stochastic recourse. The SP finds an optimal sequence in which to visit customers who are willing to pay the price, by maximizing expected revenue minus travel cost. The SP generates monotone feasibility and optimality cuts (Sect. 3.2.3).
- **Time-dependent multiple vehicle routing** [48]. The solution algorithm is a branch-and-check method (Sect. 4.2) in which the MP selects routes by selecting arcs. The SP takes into account the time required to traverse the arcs, which depends on time of day.

The SP decouples by vehicle and minimizes the return time of vehicle, which is a factor in MP objective function. Simple monotone optimality and feasibility cuts (Sect. 3.2.3) are generated.

- **Multiple vehicle routing with time windows** [86]. Standard LBBD is applied, as well as branch and check (Sect. 4.2). The MP assigns customers to vehicles and contains a relaxation of the SP (Sect. 4.6). The SP decomposes by vehicle and finds a route for each vehicle that observes time windows. Each arc is associated with a cost, which could be travel distance or some other measure, and the objective is to minimize total cost. Simple monotone feasibility cuts (Sect. 3.2.3) are used. An optimality cut is also proposed, but it seems to be invalid. The cut is

$$z \geqslant v^*(\bar{x}) - \sum_{j=1}^{n} 2(1 - x_j) \max_{k \neq j}\{c_{jk}\} \tag{5.1}$$

where $x_j = 1$ when customer j is assigned to the vehicle, c_{jk} is the cost on arc (j, k), and $\bar{x} = (\bar{x}_1, \ldots, \bar{x}_n) = (1, \ldots, 1)$ is the current solution of the MP. It is assumed that arc costs and arc travel times are symmetric and satisfy the triangle inequality. A counterexample to (5.1) appears in Fig. 5.1. A depot can be added to the example by giving it a time window $[0, \infty]$, and adding arcs from the depot to other nodes have zero cost and travel time. There is only one feasible path, namely $(5, 1, 3, 2, 4)$, which has cost $v^*(\bar{x}) = 6$. However, if $x = (1, 1, 1, 1, 0)$, a feasible (and optimal) path is $(1, 2, 3, 4)$ with cost $z = 3$. The cut (5.1) states that $z \geqslant v^*(\bar{x}) - 2 \max\{1, 1, 1, 1\} = 4$, which does not hold. However, cut (5.1) is clearly valid if there are no time windows, since an additional customer can always be accommodated by inserting the customer between two existing customers in the vehicle's tour (Sect. 3.3). The cost of doing so is at most double the maximum cost of traveling between two customers.

- **Scheduling and routing of automated guided vehicles** [61, 198, 215, 216]. In [61], the MP assigns transport requests to AGVs and schedules pickup and delivery based on shortest-path routes. The SP tries to find collision-free routes. Simple nogood feasibility cuts (Sect. 3.2) require changing the starting time of a pickup/delivery task or changing the assignment of transportation requests to AGVs. Similar approaches are used in [215, 216]. • In [198], AGVs transport items from one machine to another in a flow shop or job shop. The MP schedules jobs on machines and assigns transport tasks to AGVs. The SP does conflict-free AGV routing. The MP with valid logical constraints is solved by Lagrangian relaxation. If the SP is infeasible, an attempt is made to construct a feasible solution. If this fails, simple feasibility cuts (Sect. 3.2) are generated.

- **Bus sightseeing** [133]. The problem is to assign tour groups to buses and route each bus to the various attractions. A branch-and-check method (Sect. 4.2) is used. The MP assigns groups to tour buses and points of attraction to each bus. The SP assigns arcs of the road network to buses, as well as start times at each point of attraction. Simple monotone feasibility cuts (Sect. 3.2.3) are expressed in terms of MP variables s_{ik}, where

Fig. 5.1 Counterexample to
cut (5.1). Time windows are
shown at nodes. Each arc (j, k)
has travel time $t_{jk} = 1$ except
where otherwise indicated. All
arc costs are 1 except where
indicated as 2

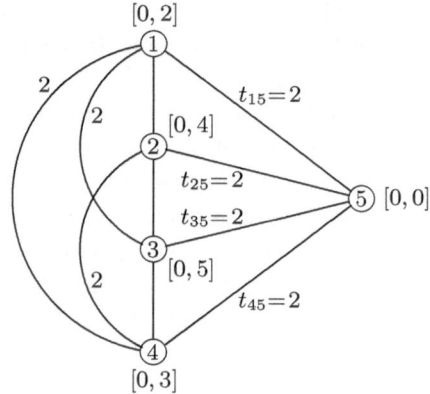

$s_{ik} = 1$ if point of attraction i is assigned to bus k. A cut is generated whenever the s_{ik}s
are all integer. A cut is also generated when the s_{ik}s are integer for some k and there is
no feasible route and schedule for bus k in SP. Thus cuts can be generated before the MP
solution is completely integer.

- **Traffic diversion** [259]. An intersection j is blocked. For all directed arcs (i, j) into the
  intersection and all arcs (j, k) out of it, the flow from i to k must be rerouted. For each
  arc e of the road network, the maximum flow through the arc subject to the capacities
  of arcs through which flow is rerouted must be within the capacity cap(e) of e. The MP
  assigns diverted flows to arcs, and the SP decouples into a flow problem for each arc. If
  maximum flow exceeds the capacity of arc e, a simple monotone feasibility cut requires
  at least one flow to go through another arc:

$$\sum_{f \mid \bar{x}_{fe}=1} (1 - x_{fe}) \geqslant 1$$

where $x_{fe} = 1$ when flow f passes through arc e, and $\bar{x}_{fe}$ is the value of x_{fe} in the MP
solution. An analytic feasibility cut

$$\sum_{f} q^*_{fe} x_{fe} \leqslant \text{cap}(e)$$

is also generated, where q^*_{fe} is the flow over arc e in the solution of the SP. There are
fewer of these than simple feasibility cuts.

5.2.2 Rail and Pipeline Transport

- **Train scheduling over individual rail blocks** [161]. A rail network is divided into stretches or blocks of track for signaling purposes, and only one train may occupy a block at a given time. The problem is to determine when each train will traverse each block on its route; each traversal is regarded as an event. The problem is solved by specifying precedences between each pair of events. Each choice of precedence is a disjunctive constraint (A precedes B or B precedes A). Some precedences are specified in the MP, in such a way that the remaining precedences decouple into subsets that can be checked independently in the SP. Thus the partition of variables depends on the problem instance. The SP is infeasible if all completions of the disjunctive choices result in a cycle in the disjunctive graph. An analytic feasibility cut is generated by identifying and excluding cycles.
- **Rescheduling trains after disruptions** [156–158, 170]. In [156, 158], the problem is to reschedule train movements dynamically as delays occur. The MP schedules trains on lines between stations. The SP decouples by station and checks whether trains within a station can be scheduled without violating station capacity. A branch-and-check method (Sect. 4.2) generates cuts for every integer node in the branch-and-cut tree that solves the MP. Analytic feasibility cuts for a given station s have the form

$$\sum_{\{i,j\} \subset K} x_s^{ij} \leqslant \tfrac{1}{2}(c_s + 1)c_s - 1$$

for all $K \subset T$ with $|K| = c_s + 1$, where $x_s^{ij} = 1$ if trains i and j meet at station s, T is the set of trains, and c_s is the number of platforms in station s. • In [157], the problem is extended to a more general traffic management problem with the structure of a job shop scheduling problem, and analytical feasibility cuts are developed. • In [170], trains are rescheduled after a massive disruption in part of the network. The MP builds a schedule for main lines. The SP decouples by station and checks for whether there is a feasible schedule at each station. Problem-specific monotone feasibility cuts require that one or more lines be removed from an infeasible station.
- **Gantry crane scheduling in a railroad container terminal** [110]. The problem and solution are related to those in [236], but with emphasis on an inland railroad terminal rather than an ocean port. Each crane can be assigned to move cargo from any train car to any truck or vice-versa, but the horizontal range of operation for any two cranes must not overlap over the entire time horizon. The MP assigns tasks to cranes, and the SP sequences tasks for each crane using CP or MILP. The objective is to minimize makespan. A simple feasibility cut excludes a solution if its makespan is worse than the previous best. Analytical optimality cuts take into account sequence-dependent setup times. One cut, similar to that in Sect. 3.4.3 but specialized to the problem, provides a bound that results from removing a task from a crane. A related cut considers the case in which a task is added to a crane (see Sect. 3.3).

- **Railway crew rostering** [199]. The problem is to construct a set of working schedules (rosters) for crews that satisfy several work regulations. A roster covers several days. The MP determines 0–1 variables f_{kd} and δ_{ik}, where $f_{kd} = 1$ when day d is the last day in roster k (rest day), and $\delta_{ik} = 1$ when duty i assigned to roster k. The SP decouples by roster and decides which duty is performed on each day of each roster. An enumerative optimality cut (Sect. 4.3) requires that some δ_{ik} or f_{kd} be flipped to 0. If a roster is infeasible, a feasibility cut requires that δ_{ik} be flipped to 0. If the SP generates a feasible roster by relaxing the number of days used, a suitable cut is generated. Symmetry cuts are used as well (Sect. 3.7).
- **Pipeline scheduling** [188]. A petroleum products pipeline transports multiple products in batches, often separated by inflated rubber balls. The batches cycle through the various products. Here, the MP defines the sequence of batches. The SP checks whether products are delivered on time. Analytic feasibility cuts say that if one or more batches are late, the sequence of batches beginning with first one that is late must be changed.

5.2.3   Maritime Transport

- **Ship routing and scheduling, one cargo per ship** [3]. Two vehicle routing models are applied to ship routing and scheduling in which each ship carries one cargo. Travel times are uncertain, and adjustable robust optimization is used (Sect. 4.5). The MP assigns origin-destination paths to each vehicle. Each scenario corresponds to a vertex of a polyhedral uncertainty set. The SP solves a set of LPs to determine whether time windows are observed in every scenario. Cuts are strengthened feasibility cuts. The strengthening method is unclear, but see Sect. 3.5.2 for strong analytical cuts that could be used.
- **Resource-constrained scheduling for maritime traffic management** [4]. Combinatorial Benders cuts (Sect. 4.4) are used to solve a traffic management problem for congested waters, with application to the port of Singapore. Optimality nogood cuts are also generated, based partly on valid constraints derived for the problem.
- **Coordination of inland vessels** [163, 164]. The problem is to coordinate the movements of inland vessels with the availability of berths and quay cranes in ports, using a closed-loop perspective. LBBD is combined with large neighborhood search. The MP uses CP to determine which terminal is visited in each segment of each ship's rotation. The SP computes waiting time for each ship. An analytic optimality cut reduces waiting time for each terminal-segment assignment removed (see Sect. 3.3). Disturbances are added to the model in [164].
- **Integrated deployment and movement planning for container ships** [187]. This complex model describes a general integrated planning problem for the deployment of vessels, routing of loaded containers, and repositioning of empty containers so as to maximize profit. Cost is not given in closed form, but is a submodular function of subsets of solution

elements. The MP solves a route generation problem, and the SB solves a fleet deployment and flow routing problem. An expert system is used to transform the MP solution into a practical routing that is sent to the SP. Classical Benders cuts (Sect. 2.4) are used when the LP relaxation of the MILP subproblem has an integer solution. Otherwise, simple monotone feasibility cuts (Sect. 3.2.3) are used.

- **Lock scheduling** [253]. A lockage is the movement of ships through one lock chamber simultaneously. The problem is to arrange ships in lockages. The MP assigns ships to lockages. The SP decouples by lockage and solves a rectangle packing problem to position ships inside the chamber. Strengthened feasibility cuts (Sect. 3.2.4) for an infeasible lockage assignment are obtained in two ways. Let S be the set of ships assigned to the lockage. (1) Generate two-variable subset cuts by first solving the SP for each subset of 2 ships in S and generating a cut for each infeasible pair. Then do the same for subsets of 3 not already covered, and so forth until no subset of a given size is feasible. (2) Generate order cuts by greedily adding ships in S in arrival order to the chamber until the SP is infeasible. Generate a cut and then remove from S all ships in the cut except the last added, and repeat. A SP relaxation (Sect. 4.6.2) is created for the MP in two ways. (1) Generate order cuts from set S of all ships, except that only the first ship added to cut (rather than all but the last) is removed from S before moving to the next cut. (2) Generate a cut for all subsets of ships whose area exceeds the area of the chamber.

5.2.4 Passenger Transport

- **Dial-a-ride assignment and routing** [217, 219]. Customers request rides from a specified pick-up to a specified drop-off point within a time window, and a fleet of vans with specified capacity are dispatched. In [217], the problem is solved by branch and check (Sect. 4.2). The MP assigns requests to vehicles to maximize the number of requests served. The SP decouples by vehicle and finds feasible tours. Monotone feasibility cuts are used and strengthened in various ways (Sect. 3.2.4). A greedy heuristic is applied once, then again in reverse order. Alternatively, all irreducible cuts are obtained by enumeration, or only the minimum length cuts among the irreducible ones are used. Symmetry cuts (Sect. 3.7) are added for vehicles with no greater capacity and no greater maximum tour length. • In [219], the MP is solved by column generation. The SP generates an analytical (lifted) cut that requires that an occupied fragment of a vehicle tour that exceeds time windows be changed by removing at least one arc. An occupied fragment is a sequence of arcs in which the vehicle has passengers.

- **Assigning customers to vehicles with drivers** [174]. Customers are assigned drivers so as to maximize revenue. Each customer requests several driving tasks requiring given processing times, which may span more than one day. If a customer is accepted, all associated tasks must be performed. Preprocessing checks whether driver's shift time is exceeded on any day, excluding driving time from one customer to another. The MP assigns drivers to customers. The SP checks for feasibility, taking into account driving time between one customer's last task and the next customer's first task on a given day. If this exceeds shift time on a given day, an analytic feasibility cut excludes assigning both these customers to the same driver. If the total shift time for a driver over the week is excessive, a cut excludes assigning the set of first and last customers over the week. If the minimum overnight break is violated, a cut excludes assigning both the last customer on one day and the first customer on next day. If a set of customers assigned on a given day does not allow for the minimum break, a cut excludes assigning this set of customers to a driver on the same day.

- **Commuter carsharing** [213]. Regular carsharing schedules for commuters are determined. The MP is formed by aggregating variables that occur in the SP. Specifically, the aggregated variables in the MP are binary variables $x_{k\ell}$, where $x_{k\ell} = 1$ if commuter request k uses transport link ℓ. The disaggregated variables in the SP are $x_{k\ell v}$, where $x_{k\ell v} = 1$ if customer request k uses vehicle v on link ℓ. The SP uses MILP to check vehicle capacities subject to constraints $\bar{x}_{k\ell} = \sum_v x_{k\ell v}$, where $\bar{x}_{k\ell}$ is the MP solution value. If the LP relaxation of the SP is infeasible, a classical Benders cut is generated. If it is feasible and the solution is integer, the problem has been solved. If it is feasible and the solution fractional, the SP is solved as an MILP. If the MILP is infeasible, a logic-based cut is generated, apparently a simple feasibility cut (Sect. 3.2).

- **Rebalancing bicycle sharing stations** [84, 151, 152]. The problem is to route one or more vehicles that convey bicycles from one sharing station to another so as to rebalance inventory at the stations. In [84], a branch-and-check method (Sect. 4.2) is used to route a single vehicle. The MP is a relaxation whose solution only partially specifies the route. The SP finds a completion of the solution, if possible. The cuts are simple feasibility cuts (Sect. 3.2). • In [151], standard LBBD is used to route multiple vehicles. The MP assigns trucks to stations. The SP decouples by truck and routes the trucks by finding a tour that alternates origin and destination stations. If a truck's route exceeds the time available, a heuristically strengthened monotone feasibility cut is generated. It is strengthened by removing an origin-destination pair of stations that are close together and checking for feasibility. • In [152], a branch-and-check version of [151] is used that includes a spanning tree relaxation of the SP in the MP (Sect. 4.6.1). The SP again decouples by truck and routes the trucks. Feasible SP solutions are cached for future iterations. Strengthened feasibility cuts are as in [151].

5.2.5   Electric Vehicles

- **Electric vehicle routing with capacitated charging stations** [94]. The MP in a branch-and-check scheme (Sect. 4.2) assigns routes to customers. A route may contain two or more charging stations with specified capacities. The SP checks whether the capacity is exceeded, perhaps after some adjustments in the time windows to allow delays at charging stations. Analytical feasibility cuts are found by seeking subsets of charging operations that violate time windows, based on energetic reasoning [18] and time window reasoning. The SP is decoupled into disconnected components of the graph formed by the routes that link charging stops. Thus if two stations have no route in common, they belong to separate components.

- **Locating charging stations for car sharing under demand uncertainty** [38]. The solution method first uses conventional Benders decomposition, then stochastic branch and check (Sects. 4.2 and 4.5), both based on MILP models. The MP decides where to locate charging stations and the car capacity of each station to maximize expected profit given uncertain demand. The SP decouples by scenario and decides which trip to assign to each request for a ride. The SP generates simple feasibility and optimality nogood cuts similar to combinatorial Benders cuts (Sect. 4.4).

- **Locating wireless charging facilities** [189]. Electric vehicles can be charged wirelessly while driving over induction coils in the roadbed. The problem is to find an optimal location of charging facilities in an urban transportation network. The MP locates the facilities. The SP checks whether their power capacities meet the power needs of each EV cluster. Simple feasibility cuts (Sect. 3.2) are used. Classical cuts based on the LP relaxation of the SP are used to strengthen the MP.

- **Factory-to-customer electric vehicle routing and scheduling** [10]. Two LBBD decompositions compute factory-to-customer routing, scheduling and recharging of electric vehicles, with inventory at the factory. The more scalable decomposition assigns vehicles to customers in the MP and decouples the SP by vehicle and time period. The method uses simple monotone feasibility cuts (Sect. 3.2.3) and analytical optimality cuts that are a special case of cut (3.17) in Sect. 3.4.3.

- **Electric bus routing and charger scheduling** [7]. Electric and diesel buses are routed subject to time windows, and a schedule is created for each charging station. The MP is defined on a graph in which each vertex represents a possible bus trip, and each edge connects a pair of trips that a bus can chain together within time windows. The MP assigns a path in the graph to each bus subject to time windows, and for each edge, assigns a charging station and charging time (possibly zero) to each electric bus assigned to that edge. The SP checks whether charging periods can be scheduled at each station. Simple monotone combinatorial Benders cuts are used in an branch-and-check framework (Sect. 4.4).

- **Charging network planning for taxi fleets** [105]. The problem is to determine a robust selection of sites for charging stations based on historical customer demand data. The MP decides where to put charging stations. The SP uses a branch-and-price algorithm to schedule charging operations and select taxi routes over various customer demand scenarios. Monotone feasibility cuts (Sect. 3.2.3) require that at least one of the currently unopened charging sites be opened.

5.3   Production

5.3.1   General Task Assignment and Scheduling

- **Task assignment and scheduling** [115, 136]. This is one of the earliest applications of
  LBBD. In [136], jobs are assigned to machines that run in parallel, and scheduled on the
  assigned machines subject to time windows. The MP uses MILP to make assignments.
  The SP decouples by machine and schedules jobs using CP. Each assignment is associated
  with a cost, and the objective is to minimize total cost. The SP is therefore a feasibility
  problem, and simple monotone feasibility cuts (Sect. 3.2.3) are generated. • In [115], a
  similar method solves the same type of problem, but in which the subproblem can be a
  multistage process scheduling problem and is solved by either CP or MILP.
- **Task assignment and cumulative scheduling** [123]. Tasks are assigned to facilities, and
  scheduled on each facility, to minimize assignment cost, makespan, or total tardiness.
  Tasks can run in parallel on each facility subject to time windows (cumulative scheduling) so long as the total rate of resources consumption is within a bound. Heuristically
  strengthened feasibility and optimality cuts (Sect. 3.2.4) are used. Relaxations of the
  subproblem similar to those in Sect. 4.6.2 are included in the MP.
- **Task assignment with sequence-dependent setup times** [246, 247]. The problem is to
  assign tasks to parallel machines so as to minimize makespan while taking sequence-
  dependent setup times into account. There are no time windows. The MP uses MILP
  to assign tasks to machines. It contains a relaxation of the SP that drops the integrality
  constraint on 0–1 sequencing variables and does not exclude subtours (Sect. 4.6.1). The
  SP minimizes makespan. It decouples by machine and is solved by a method specialized
  to the traveling salesman problem. Analytical optimality cuts are generated that are a
  weaker form of the cut provided for the minimum makespan problem in Sect. 3.4.3.
- **Task assignment and scheduling with time windows and sequence-dependent setup
  times** [97]. The MP assigns jobs to machines. The SP uses CP to schedule jobs assigned on
  each machine subject to job availability intervals. Monotone feasibility cuts (Sect. 3.2.3)
  are used. When a cut is valid on one machine, it is duplicated for all slower machines
  (Sect. 3.7).
- **Task assignment and scheduling with delay-dependent processing times** [258]. In
  this problem, jobs deteriorate if their start time is delayed, and their processing time
  increases accordingly. The increase is a step function of delay. The objective is to minimize a weighted sum of total tardiness and energy consumption. In the MP, binary
  variables $y_{jk} = 1$ when job j is assigned to machine k, $z_j = 1$ when job j starts before
  its deteriorating threshold (when the processing time goes up), and $u_j = 1$ when job j is
  maintained by holding equipment. The SP decouples by machine and sequences the jobs.
  A simple monotone optimality cut containing the y_{jk}'s is used. In addition, an invalid
  "greedy assignment cut" is added to the MP, deliberately sacrificing optimality to speed
  the solution. The degree of sacrifice is assessed experimentally.

- **Task assignment and scheduling with working-time restrictions** [87]. Workers process tasks on parallel machines subject to constraints on the maximum total and consecutive working time and the break time for each worker. The objective is to minimize a weighted sum of makespan, machine depreciation costs, and labor costs. The problem is solved iteratively over increasing and decreasing time horizons. Simple nogood feasibility cuts (Sect. 3.2) are used. Additional cuts are added if the number of breaks for a given employee must be increased.
- **Task assignment and scheduling with order acceptance** [193, 254, 255]. In [193], both standard LBBD and a branch-relax-and-check procedure are used. The MP selects orders and assigns them to one of several parallel machines to maximize profit. The SP decouples by machines and sequences and schedules orders to minimize total tardiness cost. Monotone optimality cuts (Sect. 3.2.3) are generated. In the branch-relax-and-check procedure, the MP incumbent solution is sent to a SP relaxation to generate temporary Benders cuts. After convergence with the SP relaxation and these cuts, an exact SP is solved and permanent BCs added. Permanent cuts are again simple monotone optimality cuts per machine. Cuts are duplicated to exploit symmetries (Sect. 3.7), as the machines are identical. • The model in [255] considers order acceptance, sequencing, setup times, and maximum availability times of machines. The objective is to minimize makespan, and the problem is solved by 3-level LBBD (Sect. 4.7). Binary variables include $z_j = 1$ if job j accepted, and $x_{ij} = 1$ if job j is assigned to machine i, and $y_{ijk} = 1$ if job k is processed immediately after job j on machine i. The MP uses z_i and x_{ij} and a continuous relaxation of 0–1 SP variables (Sect. 4.6.1), and it passes acceptance decisions $\bar{z}_i$ to the SP. The SP is itself solved by LBBD and decomposes into MP2 and SP2. MP2 uses the same variables as MP and passes assignment decisions $\bar{x}_{ij}$ to SP2. SP2 sequences jobs on each machine and generates cuts for MP2 that are used in [246]. SP2 also generates simple monotone feasibility cuts for MP2 because it can be infeasible due to machine time limits. Cuts for MP include simple monotone feasibility and enumerative feasibility cuts (Sect. 4.3). A similar model is used in [254].
- **Task assignment and pre-emptive priority scheduling** [120]. The MP allocates tasks to processors using CP, and the SP schedules the tasks on each processor. Monotone feasibility cuts are used, strengthened by applying the QuickXplain algorithm to find a minimal infeasible set of constraints in the SP (Sect. 3.2.4).
- **Stochastic task assignment and scheduling** [75, 166]. In [75], stochastic LBBD (4.5) and branch-and-check methods (Sect. 4.5) are applied to a generic two-stage assignment and scheduling problem with integer recourse. The MP assigns tasks to agents, and the SP schedules the tasks for each agent subject to one or more uncertain parameters, such as processing time. The SP decouples by scenario as well as by agent. Analytical cuts similar to those in Sect. 3.4 are generated. A computational comparison with the integer L-shaped method is very favorable. The same method can be used for robust optimization. • In [166], stochastic branch and check with similar cuts is applied to a related problem

in which orders are assigned to factories. The model accounts for uncertain supply chain delays as well as uncertain processing times.

- **Machine location, assignment, and scheduling** [169]. The problem is to select locations for machines, assign jobs from different locations to the machines, and sequence the assigned jobs on each machine. A branch-and-check method (Sect. 4.2) is used to minimize makespan. The MP determines machine locations and job assignments to machines, while the SP sequences jobs on each machine. A monotone feasibility cut is added if the SP makespan is worse than the incumbent solution in the branching search. A deletion filter is used to strengthen the cuts (Sect. 3.2.4). Lifted cuts are obtained by solving an MILP to determine whether jobs can be added to the left-hand side of the cut. Strengthened monotone optimality cuts are also used. Analytic optimality cuts are obtained by subtracting the sum of release time and processing time from the makespan for each job removed from an assignment (see Sect. 3.4.3).

- **Assignment and scheduling of single- and multi-stage processes** [175]. Single-stage and multi-stage jobs are assigned to machines, subject to time windows for each job. In the single-stage problem, the MP assigns jobs to machines, and the SP schedules them. The MP is enhanced with time-window relaxations involving all subsets of jobs (see Sect. 4.6.2). There is a stronger constraint when the same job has the earliest release time and the latest deadline. Simple monotone feasibility cuts are used when minimizing cost, and simple monotone optimality cuts when minimizing makespan. In the multistate problem, the SP is a job shop problem that does not decouple by machine. The SP is solved by a shifting bottleneck algorithm, and simple monotone feasibility cuts are generated. If the jobs assigned to an individual machine cannot be scheduled, a cut is generated for that machine; these jobs are identified during the shifting bottleneck procedure. Otherwise, a cut is generated for the overall SP.

- **Machine assignment and scheduling with job splitting** [11]. Jobs with sequence-dependent setup times are assigned and scheduled so as to minimize makespan, with no time windows. The MP splits jobs into pieces (defined by percent of time required) and assigns pieces to machines. The SP uses constraint programming and generates simple monotone optimality cuts (Sect. 3.2.3).

- **Computational analysis of LBBD for facility assignment and scheduling** [57]. The aim is to identify factors that reduce LBBD solution time in this class of problems. It is observed that the SP and MP should consume about equal time, the number of iterations should be less than 100 or so, and it is important to use a SP relaxation in the MP (4.6).

5.3.2 Shop Scheduling

- **Job shop scheduling with sequence-dependent setup times and worker time windows** [153]. The problem is solved by a branch-and-check method that sends each integer solution obtained to the SP. The MP assigns jobs to machines and sequences jobs on each machine. The MP variables $x_{ij}^m = 1$ when job i immediately precedes job j on machine m. The SP schedules workers on each machine. Monotone optimality cuts (Sect. 3.2.3) are generated by requiring that the objective function (makespan or tardiness) is the same unless some x_{ij}^m is switched from 1 to 0.

- **Flexible job shop scheduling** [143, 194]. The job shop is flexible in the sense that there is some choice as to which machine an operation is assigned. The objective is to minimize makespan, with no time windows. In [143], the MP assigns each operation of each job to an eligible machine. The SP sequences and schedules the operations assigned to each machine. The SP does not decouple due to precedence relations. After the MP is solved, a relaxed SP is solved. The relaxed SP is based on bounds obtained from the running times of operations. If the solution is no better than the incumbent best makespan, a strengthened optimality cut (Sect. 3.2.4) is generated. Otherwise the complete SP is solved, and a simple monotone optimality cut obtained. • In [194], the MP and SP divide up tasks as in [143]. Precedence relations and processing times are used to add a relaxation of the SP into the MP (Sect. 4.6.2). The minimum makespan subject to these data is formulated as an MILP, which is incorporated into the MP by introducing 0–1 variables. Analytic optimality cuts reduce the makespan bound for each operation-machine assignment that is removed (Sect. 3.4.3). Then, another version of the SP is solved to allow a change of machine assignment for operations that are on the critical path. This new assignment is sent to the original SP. The method cycles through this maneuver three times, generating three additional cuts.

- **Flowshop scheduling** [111–113, 121, 201, 240]. In [111–113], each job is processed on machines in a fixed order, with a given specified time lag between consecutive processing stages. The objective is to minimize the number of late jobs. The time horizon divided into intervals, and a job's processing time on a machine cannot overlap 2 intervals. The MP assigns jobs to time segments, and the SP schedules jobs within each interval. Simple monotone optimality cuts (Sect. 3.2.3) are used. • In [113], an upper bound on the completion time of each job is obtained. • In [121], there are time-of-day energy costs. The MP sequences jobs in a flowshop with two machines. The SP finds start times that minimize electricity cost, where the electric rate charge depends on the time of day. Simple monotone optimality cuts (Sect. 3.2.3) are used. • In [201], combinatorial Benders cuts (Sect. 4.4) are used to schedule flowshops with constraints on the queuing time between consecutive steps. • In [240], there are two processing stages. A single machine is available for stage 1 and unrelated parallel machines for stage 2. The objective is to minimize makespan. The MP variables are $x_{kijj'} = 1$ if job j precedes job j' on machine i in stage $k \in \{1, 2\}$, and $v_{kij} = 1$ if job j is assigned to machine i in stage k. A monotone optimality cut is used:

$$z \geqslant v^* \left(1 - \sum_{j | \bar{v}_{2ij}=1} (1 - v_{2ij}) - \sum_{j | \bar{x}_{11jj'}=1} (1 - x_{11jj'}) \right)$$

where $\bar{x}$, $\bar{v}$ are the MP solution values, and v^* is the optimal makespan in the subproblem. In addition, the MP contains a relaxation of the SP (Sect. 4.6).

- **Single-machine scheduling to minimize number of late jobs** [225]. A branch-and-check algorithm minimizes a weighted sum of late jobs on a single machine with release dates. The MP contains only the objective function, cuts, and a relaxation of SP consisting of the inequality

$$\sum_{\ell} \min \left\{ d_j - r_i, \ \left(p_\ell - \max \left\{ (r_i - r_\ell)^+, (d_\ell - d_j)^+ \right\} \right)^+ \right\} x_\ell \leqslant d_j - r_i$$

for each time window $[r_i, d_j]$. Here $\alpha^+ = \max\{\alpha, 0\}$, and $x_\ell = 1$ means that job ℓ is on time. A monotone feasibility cut (Sect. 3.2.3) is strengthened by using a modified Carlier algorithm to find a small infeasible subset of jobs (i.e., a subset of jobs that cannot all be on time). Analytical strengthening is obtained by an application of edge finding before including the above SP relaxation in the MP, on the assumption that some subset of jobs are scheduled on time.

- **Single-machine scheduling over a long time horizon** [59]. The aim is to solve single-machine scheduling problems with a long time horizon and many jobs, by breaking the time horizon into segments of manageable size. The MP assigns jobs to segments, and the SP uses CP to schedule the jobs subject to time windows. One version assumes that a job's processing period cannot extend over two consecutive segments, perhaps because they are separated by a weekend. Heuristically strengthened feasibility and optimality cuts (Sect. 3.2.4) are used to minimize makespan or total tardiness. Another version allows a job to overlap two or more segments, which requires additional MP variables and more complex cuts.

5.3.3   Assembly Line and Work Cell Management

- **Assembly line balancing** [95, 183, 226, 230, 269]. In [95], the MP assigns tasks to work stations. The SP decouples by work station and sequences tasks on each work station using sequence-dependent setup times. Simple monotone feasibility cuts (Sect. 3.2.3) require that a task be removed from an infeasible work station. • In [183], multiple workers can be assigned to a work station. The MP assigns tasks to stations and workers to stations. To relax the SP in the MP, infeasible pairs of tasks are identified in the MP, based on a precedence graph and processing times (4.6.1). The SP schedules tasks for each worker and checks feasibility at each station. A feasibility cut requires that either a task be removed from the station, or at least one worker be added to the station. The cut simplifies if a station already has the maximum number of workers. • In [226], the problem has a dual objective of minimizing the number of workers and the number of work stations. Workers can walk from one station to another. The MP assigns tasks to workers to minimize the number of workers. The SP minimizes the number of work stations by assigning tasks to stations and determining their start times. The LBBD procedure first minimizes the number of workers by checking SP feasibility and adding simple monotone feasibility cuts. The minimum number of workers in this solution is noted. The procedure then keeps adding feasibility cuts in an attempt to reduce the number of work stations. It terminates whenever the number of work stations is not reduced, or the number of workers is increased (or the number of work stations reaches known lower bound). A primal heuristic is integrated into the procedure. • In [230], demand is stochastic, and the problem is solved by branch and check. The MP assigns tasks to stations. The SP requires that jobs visit each station in the same order and measures setup inefficiencies. An incomplete set of analytical cuts is obtained by considering subsets of stations and imposing the resulting simple monotone optimality cuts. This can result in useful cuts because they contain fewer variables. The implementation selects groups of up to 3 stations, not all combinations but enough to cover all the stations. • In [269], the sequencing SP accounts for setup times. Simple monotone feasibility cuts are used (Sect. 3.2.3) in a branch-and-check (Sect. 4.2) context, along with optimality cuts that are a special case of the makespan cut described in Sect. 3.4.3. Additional analytic cuts exclude certain specific task assignments by setting the corresponding 0–1 variables to zero.

- *Seru* (**work cell**) **assignment and scheduling, with learning** [268]. A *seru* is a type of work cell originating in the Japanese electronics industry. Each *seru* assembles products to completion, with each worker staffing two or more stations. This model incorporates DeJong learning curves. The MP assigns jobs to *serus* by defining binary variables x_{ij}, where $x_{ij} = 1$ when job j assigned to *seru* i. The MP also contains 0–1 sequencing variables that are relaxed to be continuous. The SP sequences jobs in each *seru* so as to minimize makespan. Each seru i generates an optimality cut that is a special case of the makespan cut in Sect. 3.4.3, namely

$$z \geqslant v^*(\bar{x}) - \sum_{j \in J_i}(1 - x_{ij})\Big(\max_{j'} \{s_{ij'j} \mid j' \in J_i \setminus \{j\}\} + p_{ij}\Big)$$

where $v^*(\bar{x})$ is the optimal value of the SP, J_i is the set of jobs assigned to *seru* i by the MP, p_{ij} is the processing time of job j in *seru* i, and $s_{ij'j}$ is the setup time between jobs j' and j in *seru* i.

- **Part assignment and transport in a robotic cell** [5]. The MP assigns parts to machines, and the SP sequences jobs and transports parts between machines to minimize makespan. Both MP and SP are MILP problems, and SP does not decouple by machine. Simple monotone optimality cuts are used (Sect. 3.2.3).

5.3.4 Employee Scheduling

- **Employee shift assignment and scheduling** [73, 214]. In [73], the MP is an MILP model that assigns workers to shifts. The SP assigns days to worker-shift pairs. Simple monotone feasibility cuts are used (Sect. 3.2.3). To speed solution, various discrete relaxations of the SP are solved first. If one of these is infeasible, we already have a valid cut. Otherwise, the SP is solved exactly. • In [214], each employee is assigned a set of activities each day (including breaks and days off) that must be scheduled subject to constraints. All feasible activity schedules ("tours") are generated for each employee, and 0–1 variable $x_{et} = 1$ if tour t is assigned to employee e. The MP solves for variables x_{et}. The SP generates schedules and decouples by day. It is formulated as a context-free grammar from which an MILP model is constructed. Classical Benders cuts and monotone optimality cuts (Sect. 3.2.3) are used.
- **Employee assignment to manage customer queues** [242, 243]. The problem is to assign employees to the front office for customer service in, say, a bank, and others to the back office. Some employees are specialized and therefore less expensive, while others are cross-trained and more expensive. Workers are switched between assignments when the customer queue reaches a certain length. The MP decides the worker assignments. The SP uses CP to formulate a switching policy to make sure waiting time is no greater than the specified limit. If the SP is infeasible, a feasibility cut requires that the number of front office workers, or back office workers, or cross-trained workers be increased.

5.3.5 Maintenance

- **Shop scheduling with machine maintenance** [17]. The problem is solved on a rolling basis, using LBBD in each period. The MP uses MILP to decide which machines to maintain in each remaining period, subject to capacity constraints. The machine speed declines with a greater lapse since last maintenance. The SP uses MILP to schedule jobs in the current period k, given the maintenance status of machines. This sacrifices optimality because jobs are not scheduled for all remaining periods. A simple nogood optimality cut (Sect. 3.2) imposes a period-k cost if exactly the same set of machines are maintained in period k. There is a SP relaxation in the MP based on the minimum completion time and maximum time available (Sect. 4.6.2).
- **Aircraft maintenance** [257]. This is an integrated inventory-location problem faced by an aerospace company in designing its service parts logistics network. Stochastic customer demand is Poisson distributed. Binary decision variables in the MP are $y_i = 1$ when service center j is open, $x_{ijk} = 1$ when customer i demand for part k is assigned to center j, and $v_{jks} = 1$ when s units of part k are stored at center j. The MP omits service level constraints. The SP simply checks the service level constraints, which are nonlinear

before the MP variables are fixed. A simple feasibility cut says that if the current stock level of part k is kept, and the current fill rate does not meet the required service level, then assignment variables x_{ijk} for this k must change. This is strengthened to an analytic feasibility cut by taking into account the impact of reassigning customers. The analysis is quite complex and problem-dependent.

- **Aircraft repair scheduling** [15, 16]. Aircraft needing repair are assigned to upcoming missions and are thereby assigned a repair deadline equal to the start time of the assigned mission. Each aircraft repair job requires one or more resources in the repair shop. The MP assigns deadlines to jobs. The SP uses CP to solve the cumulative scheduling problem (Sect. 3.4.1) subject to deadlines. If there is no feasible schedule for a given resource r, an analytical optimality cut (Sect. 3.4) peculiar to this problem is generated as follows. Let binary variable $x_{ij} = 1$ when deadline i is assigned to job j. Then the cut says that at least one of the jobs using resource r must have a later deadline than currently assigned. The cut is

$$\sum_j \sum_{i \in I_j} (1 - x_{ij}) \geqslant 1$$

where j ranges over the jobs using resource r and I_j is the index set of deadlines that are on or before the deadline currently assigned to job j. The MP contains a SP relaxation similar to those in (Sect. 4.6.2).

- **Wind turbine maintenance** [93]. Maintenance operations for onshore wind turbines are scheduled so as to maximize electricity production while taking into account expected wind speeds and daily restrictions on how technicians are assigned and routed. The MP schedules maintenance tasks, and the SP assigns technicians to tasks. A branch-and-check framework (Sect. 4.2) is used with simple feasibility cuts plus highly specialized analytical cuts that exclude assignments that create infeasibility in either of two relaxations of the SP.

- **Location/allocation of agricultural machinery maintenance** [114]. The MP assigns maintenance facilities to locations and demand points to facilities, and a linear SP checks for contiguity of the area served by each facility. The SP decouples by service facility and time period. Simple monotone feasibility cuts (Sect. 3.2.3) are used, apparently in a branch-and-check algorithm (Sect. 4.2). A warm start is obtained by solving the subproblem for an individual time period and checking feasibility.

5.3.6 Factory Scheduling and Production Planning

- **Integrated production planning** [180, 181, 190]. In [181], branch and check (Sect. 4.2) determines optimal process configurations, lot sizes, and product sequencing simultaneously. The MP estimates the quantity q_{is} of each product i to be made in period s. The SP solves the exact problem. If the SP is infeasible, an analytic feasibility cut imposes a bound on q_{is}. In particular, the current value of q_{is} is reduced by an appropriate amount if the process configuration is changed to achieve feasibility. The cut is rather complex due to the complexity of the model. • A similar solution method is used in [180] for integrated production planning. Here, a SP relaxation is added to the MP, consisting of valid inequalities for subsets of products with similar characteristics (Sect. 4.6.2). The method is applied to cutting stock problems and a printing problem. • In [190], each job is associated with a directed network showing how operations and parts must be organized to create a product. The MP decides how products are assembled based on a processing and/or graph that indicates product flow through processes. Each node corresponds to a stage in the processing of a particular product. The MP assigns machines to nodes and determines which path is taken out of each OR-node. In particular, 0–1 variables $w_{ijk} = 1$ if operation k of job j is processed on machine i, an assignment that requires processing time p_{ijk}. The SP does the sequencing and scheduling. The objective is to minimize makespan. The MP contains a relaxation of the SP (Sect. 4.6) based on the processing network. Analytic optimality cuts reduce the makespan bound by p_{ijk} for each w_{ijk} switched to 0 (see Sect. 3.4.3).

- **Robust production scheduling with energy consumption limits** [185]. Branch and check (among other methods) is applied to the problem of finding an optimal robust schedule of production tasks that continuously satisfies a limit on the rate of energy consumption. The energy cost depends on the maximum consumption rate, which is checked every 15 minutes or so. The MP finds a baseline sequence and schedule of tasks that satisfies the energy consumption limit. The SP computes an optimal robust schedule that takes into account worst-case delays, given the sequence from the MP. Let $\bar{s}_1, \ldots, \bar{s}_n$ be the start times of the jobs in the optimal baseline sequence. If the SP is infeasible, an analytic feasibility cut is generated. It is based on the fact that if every job j starts within the interval $[\bar{s}_j, \bar{s}_{j+1} + p_{j+1})$, where p_j is the processing time of job j, then the optimal sequence must be the same as before. Therefore, the cut requires that at least one s_j lie outside this interval. More complex cuts of this kind are generated when the SP is feasible but delivers an optimal robust schedule that is different from the baseline schedule.

- **Blast furnace "torpedo" scheduling** [106]. A torpedo is a car that moves molten metal from a blast furnace to a desulfurizer and oxygen converter, and then returns to the blast furnace empty. The model is based on blast furnace events (release of molten metal) and oxygen converter events. The MP solves an MILP to assign torpedo runs to converter events (whereupon the torpedo returns to the furnace) using a 2-criterion lexicographic objective: first minimize the number of torpedo runs, then minimize a lower bound on desulfurization time. The SP schedules batches on the converter. The SP is solved by CP and decouples into subsets of torpedo runs according to which ones can have interfering space-time trajectories, based on the time windows. Nogood feasibility cuts (Sect. 3.2) require that at least one torpedo run be assigned to a different converter event:

$$\sum_i (1 - x_{i\bar{x}_i}) \geqslant 1$$

where $x_{ij} = 1$ when converter event j is assigned to blast furnace event i, and $\bar{x}_i$ is converter event assigned to blast furnace event i in the current MP solution. The cuts are strengthened using a deletion filter (Sect. 3.2.4). If the desulfurization time is Δt less than the lower bound computed in the MP solution, a penalty term $\delta \Delta t$ is added to the MP objective function, where δ is a new 0–1 variable defined by adding the inequality

$$\sum_j (1 - x_{i\bar{x}_i}) \geqslant 1 - \delta$$

to the MP constraint set.

- **Raw material handling in the steel industry** [235]. The problem is to manage transfer of raw materials to steel mills. The MP assigns jobs to "reclaimers," which are machines that take raw material from a pile. The MP sequences jobs on each reclaimer and on each conveyer to the factory. The SP is an LP model that checks for overlap and time windows. Combinatorial Benders cuts are used (Sect. 4.4).
- **Scheduling steel manufacture** [139]. In a multi-stage steel making process, the last stage is continuous casting in which certain consecutive batches must be contiguous in time. The MP determines a schedule for the last stage, assuming it will not be delayed by operations in previous stages. The SP determines a schedule for the previous stages, given the last stage schedule and any unexpected delays in the previous stages. If there is no way to accommodate the last stage schedule, an analytic feasibility cut requires that at least one job in the last stage that overlaps a previous job be postponed by an amount equal to how much it overlaps the previous job.
- **Parts assembly planning** [129]. The problem is to find a sequence in which parts are assembled so as to minimize some measure of cost while making sure that there is a fixture to which each part can be attached in the current subassembly, and so that there is room to manipulate the tool that attaches the part. A sequence is viewed as a series of tasks, each of which consists of using a certain tool and fixture to attach a specified part to another specified part on the subassembly. The MP uses MIP to find an optimal sequence, and the SP checks whether the sequence is physically possible. If a particular task results in a collision, an analytic feasibility cut requires that the task not be performed, or if it is performed, one of the parts, the tool, or the fixture not be part of the task.
- **Allocation of software components to processors** [135]. The problem is similar to computer processor scheduling but applied to manufacturing control. The MP assigns software components to processors and the SP schedules operations under a worst-case scenario. Strengthened feasibility cuts are obtained using QuickXplain (Sect. 3.2.4).
- **Batch scheduling and lot sizing in the chemical industry** [245]. This is one of the earliest applications of branch and check (Sect. 4.2), or a variation thereof. The problem is to manage a multistage chemical processing plant over multiple periods by determining lot

sizes and scheduling tasks on each machine. The MP solves a dynamic lot-sizing problem by MILP and, in particular, determines which product is assigned to each machine in each time period. The SP determines whether the tasks can be scheduled, taking into account sequence-dependent costs. After every few nodes are enumerated in the MILP branching tree, the assignment variables that have been fixed so far are sent to a relaxed SP to check for feasibility. If the SP is infeasible, a simple feasibility cut is generated that loses validity upon backtracking. If the relaxed SP is feasible, a CP model of the original SP is invoked, and a simple feasibility cut generated if it is infeasible.

- **Assignment and scheduling of batch processes** [176, 177]. The problem is to schedule operations in batch processing plants that can be given a state task network representation. The MP uses MILP to decide which processing tasks to perform, assign them to units, determine batch sizes, and enforce mass balance constraints without reference to time periods. The SP uses CP to schedule tasks. Three types of cuts are used. One is a simple feasibility cut (Sect. 3.2). Another is a stronger cut based on infeasibility detected in a component of the state task network. A third type of cut is derived from the previous cuts due to the similarly of tasks (Sect. 3.7).

- **SAT modulo theories for chemical process engineering** [184]. This paper shows how chemical process engineering problems can often be modeled as SAT (propositional satisfiability) modulo theories. It points out the connection between finding an unsatisfiable core of the theory component and finding strengthened logic-based Benders cuts. In fact, the conflict clauses obtained by the theory solver are logic-based Benders cuts, and SATM solvers are special cases of branch and check.

- **Underground mine scheduling** [171]. LBBD is combined with a heuristic to schedule tasks involved in the evacuation of an underground cut-and-fill mine, taking into account priorities for different locations in the mine. The MP uses MILP to assign tasks to shifts and machines to tasks. The SP uses CP to schedule tasks within a shift, based on machine assignments. Strengthened monotone feasibility cuts are obtained for infeasible shifts by greedy cut strengthening and a deletion filter (Sect. 3.2.4).

5.4   Supply Chains

5.4.1   Upstream Logistics

- **Space allocation in an inland bulk material stockyard** [234, 241, 251]. The problem is similar to a container stacking problem described under "Managing container stacking and relocation in a seaport" in Sect. 5.4.3, but here one is storing bulk material such as iron ore or coal at an inland stockyard. The material is stored on pads that are divided into slots. The MP determines binary variables u_{iw} to indicate whether task i is unloaded at unloading station w, x_{ik} to indicate whether incoming task i occurs after reclaiming task n and before reclaiming task $n + 1$ on stock pad k, and y_{ik} to indicate whether incoming material i allocated to stock pad k. The SP determines binary variables z_{iks} to indicate whether stacking of incoming material i ends at slot s in stock pad k, in which case it allocates space on the pad. An SP relaxation is included in the MP by dropping integrality restrictions on z_{iks} (Sect. 4.6.1). The SP is an MILP problem that decouples by pad. Decoupled simple monotone feasibility and optimality cuts (Sect. 3.2.3) are used for variable y_{ik}. In addition, if I_k is the set of materials assigned to pad k, an analytical optimality cut is obtained by considering each material $j \notin I_k$ and solving an easy problem to determine the minimum cost γ_{jk} of adding material j to pad k on top of I_k (see Sect. 3.3). The cut is

$$z \geqslant v^* - v^* \sum_{i \in I_k}(1 - y_{ik}) + \sum_{i \notin I_k}\gamma_{ik}y_{ik}$$

where v^* is optimal value of the SP. Similar approaches are described in [241, 251].

- **Container drayage** [34, 260]. In [34], the problem is to assign and schedule trucks at a single depot to pick up and deliver containers (possibly empty) to/from customers over multiple periods (days). The MP assigns customers to subtours (trips) in the transport network. The SP, a feasibility problem, assigns trucks and periods to trips. Of several types of cuts tried, the best results are obtained from simple monotone feasibility cuts (Sect. 3.2.3) and a SP relaxation in the MP (Sect. 4.6) that requires the total duration of trips chosen in a given period to be no greater than the number of trucks times the max allowed travel time per truck. • In [260], tractors and trailers are scheduled separately. Tractors can be assigned to a new task at another location while trailers with containers are waiting for packing or unpacking. Every pickup request is represented by an empty-pickup node and a loaded-pickup node in a graph, in which there is also a terminal node. A tractor hauls an empty trailer to the empty-pickup node, and a possibly different tractor hauls the container to the terminal after it has been loaded. Similar delivery nodes are defined for every delivery request. The MP minimizes cost by determining variables x_{ij}, where $x_{ij} = 1$ when task nodes i, j are served consecutively by the same tractor. The SP is an LP problem that determines start times of tasks so that empty delivery precedes loaded pickup by enough time to allow loading, etc. The problem is solved by branch and check using combinatorial Benders cuts (Sect. 4.4).

- **Cross-dock goods transfer scheduling** [47]. Cross-docking refers to the transfer of goods directly from an arriving vehicle to a departing vehicle parked at a cross dock, without warehouse storage and order picking. This model considers the 3-D shape of boxes transferred. The MP assigns shipments to cross docks, and the SP assigns boxes to trucks. The SP decouples by origin and destination. Let z_{fd} be a bound on the cost of moving cargo through dock f to destination d, and let v^*_{fd} be the value obtained for this cost in the optimal solution of the SP. Also let δ_{kf} be a binary MP variable that is 1 when shipment k is assigned to dock f. The following analytical optimization cut is generated for all f, d, k:

$$z_{fd} \geqslant v^*_{fd} - \sum_k M_{kfd}(1 - \delta_{kf})$$

where M_{kfd} is an upper bound on the cost of routing shipment k through f to d.

- **Two-echelon distribution center location and vehicle routing under uncertain demand** [186]. The problem is to locate warehouses and the fulfillment centers they supply, and to route vehicles that transfer items from one to the other, under uncertain demand. It is modeled as a two-stage stochastic program with integer recourse and solved by stochastic LBBD (Sect. 4.5). The MP makes location and capacity decisions for warehouses and fulfillment centers. The subproblem uses branch and price to solve a capacitated vehicle routing problem after observing stochastic demands. The SP decouples by time period and scenario. Classical Benders cuts are obtained from the LP relaxation of the SP, and logic-based cuts are obtained from the SP solution. Optimality cuts say that if the capacity of every facility in the current time period is no greater than in current MP solution, then the cost incurred in each scenario in the current time period is at least the cost incurred in the current SP solution. Since the cut is conditional in form, 0–1 variables are introduced to linearize the cut.

- **Lot sizing and vehicle routing to transport parts for assembly** [2]. Capacitated vehicles carry parts from a supplier to an assembly plant. The objective is to minimize cost, where demand and all costs vary over time. LBBD is applied to a problem with a single vehicle that visits all suppliers. The MP solves a dynamic assembly lot sizing problem by MILP. The SP decouples by period and solves a traveling salesman problem using Concorde. The method uses simple monotone optimality cuts (Sect. 3.2.3). In this case, LBBD is not the fastest method tried. Branch and check could have better performance but is not applied.

- **Multi-depot product delivery with transshipments** [265]. This model is primarily intended for "last mile" delivery logistics in which products are transferred to local delivery vehicles at intermediate terminals. It is solved with a branch-and-check method using combinatorial Benders cuts (Sect. 4.2). The SP consists of two LP problems, one of which checks time windows and the other vehicle capacity.

- **Plant location with customer/truck allocation** [88–90]. The problem is to select sites for plants and assign customers to plants as well as to delivery trucks, subject to driving distance constraints. In [88, 89], the MP uses MILP to select facility locations, assign customers to facilities, and decide how many trucks are allocated to each facility. The SP uses CP to find the minimum number of trucks needed to served customers assigned to each facility (bin packing) and decouples by facility. Trucks make a round trip for each customer and can make several round trips in a day if total driving distance is within bounds. An analytical optimality cut says that the number of trucks that must be assigned to a facility is at least the minimum computed in the SP minus the number of customers removed from this facility (each customer requires a separate truck in the worst case, as illustrated in Sect. 3.3). A small error is corrected in Sect. 4.4.3 of [222]. • A version of the problem with uncertain travel times is solved by stochastic LBBD (Sect. 4.5) in

[90]. A penalty is imposed for each truck that must work overtime. An assignment of customers to a specific truck at a facility is infeasible if the expected penalty exceeds Q. An analytic optimality cut says that the expected penalty must be at most Q, where the penalty in each scenario is the current penalty minus the maximum reduction in penalty when a customer is unassigned from the truck (see Sect. 3.3).

- **Concrete delivery to construction sites** [150]. The problem is to schedule concrete deliveries to various construction sites, taking into account that it is often necessary for deliveries to a given site to arrive in close succession. The MP uses MIP to select customers to whom to deliver concrete. The SP uses CP to schedule deliveries. Monotone feasibility cuts (Sect. 3.2.3) are strengthened heuristically by gradually adding customers to the subproblem until infeasibility is encountered (see Sect. 3.2.4 for other options). A warm start is obtained by enumerating all irreducibly infeasible subsets of up to three customers, as well as by sending heuristically generated customer assignments to the subproblem.

- **Wheat supply chain management with capacity decisions** [192]. The problem is to install capacity in distribution centers (DCs) and manage the movement of wheat from suppliers to DCs to customers. The MP selects which DC center to open, assigns discrete capacities to the centers, assigns customers and suppliers to DCs, and estimates the number of transporters needed. The SP decouples by DC center and assigns transporters to suppliers and customers. The problem is solved if the SPis solved to optimality for all DCs, and the number of transporters is equal to that in the MP solution. A simple optimality cut (Sect. 3.2) requires the number of transporters to be the same as in the MP unless, for some DC in which the number of transporters is greater than in the MP solution, some supplier or customer is unassigned from the DC. If the SP is infeasible for some DC, a simple monotone feasibility cut (Sect. 3.2.3) is generated for that DC. If the total number of transporters exceeds availability, an overall simple monotone feasibility cut is generated.

- **Intermodal transport from shipper to customer** [12]. The problem is to ship orders from pickup points to depots in Europe, from depots to warehouses in another country by various modes, and from warehouses to customers. The MP is an MILP model that determines assignment of orders to warehouses and to modes. The SP schedules deliveries from warehouses to customers, given a limited number of trucks in each warehouse and due dates. The objective is to minimize the sum over orders of tardiness squared. The SP does not decouple and generates simple monotone feasibility and optimality cuts (Sect. 3.2.3). The MP includes a time window relaxation of the SP (Sect. 4.6.2) that is specialized to this objective.

- **Supply chain reconfiguration** [202]. A goal programming model is formulated to determine how to change outsourcing strategies to meet increased customer demand and requirements in the aerospace industry. The SP is an LP problem, and classical Benders cuts are used when the SP is feasible. Otherwise, combinatorial Benders cuts (Sect. 4.4) require that at least one change be made in either the set of tier 1 suppliers or the set of tier 2 suppliers.

5.4.2 Warehousing and Inventory

- **Order picking in a warehouse** [72]. The problem is to route an order picker in a U-shaped storage area. The picker makes several trips to shelves to collect items that are brought to a movable depot. The MP minimizes the travel distance of the picker, with a lower bound computed geometrically. It selects which items are collected on each trip, and how many trips. It includes a relaxation of the SP (Sect. 4.6) based on the circumference of the convex hull of shelves visited by picker on each trip. This circumference is approximated by a Manhattan metric and Chebyshev distance. The SP defines the location of the depot and the route of the picker for each trip. The routing problem becomes trivial once a location is selected for the depot. An analytic feasibility cut requires that at least one item be removed from at least one trip. It is strengthened by noting that only visited shelves on vertices of the convex hull of visited shelves need be considered. Removing other items has no effect on walking distance, because the picker passes by that shelf anyway.

- **Robotic repositioning of pods in a fulfillment center** [231]. This paper addresses fulfillment centers in which mobile robots bring movable inventory pods to stationary human pickers at fixed stations. It gives an exact LBBD method for a stochastic problem, but uses a heuristic for realistic instances. The MP is an integrated picking problem for a given "wave" of orders. The SP considers a set of scenarios for the next wave of orders, while receiving pod locations left over from the previous wave as specified by the MP. Simple optimality cuts include the expected cost in next wave resulting from these pod locations.

- **Safety stock management** [203]. There are multiple stockpoints from which materials are sourced, and the objective is to minimize holding cost. The MP solution determines the value $\bar{x}_{ijk}$ of x_{ijk}, where $x_{ijk} = 1$ if stockpoint j is used to hold the safety amount of item k at stage i in the supply chain. The SP is a continuous nonlinear programming problem that regulates stock levels, and checks whether they exceed stockpoint capacity. A simple feasibility cut (Sect. 3.2) is used if the SP is infeasible. Otherwise, an analytical optimality cut is generated that bounds the holding cost by the sum of the optimal SP value and $\sum_{ijk} \lambda_{ijk}(x_{ijk} - \bar{x}_{ijk})$, where λ_{ijk} is the change in the SP objective when x_{ijk} flips from 0 to 1 (see Sect. 3.3). To compute λ_{ijk}, it is noted that $\sum_j x_{ijk} = 1$ for all i, k. The SP is solved for all $j' \neq j$ to note the change in optimal value when $x_{ij'k}$ is set to 1 rather than x_{ijk} set to 1. There seems to be a typo in formula (30).

- **Shelf space allocation** [98]. The MP decides the vertical and horizontal dimensions of space allocated to each product on shelves, as well as the number of units of each product displayed in each cabinet (which is implied by the dimensions). The SP arranges the products in each cabinet. Simple monotone feasibility and optimality cuts (Sect. 3.2.3) are used. The optimality cuts are valid but are written in an unusual fashion.

5.4.3 Container Port Management

- **Scheduling trucks and yard cranes in a container port** [44]. The MP uses MILP to determine assignment of jobs to trucks and yard cranes, and sequencing of the jobs. Combinatorial Benders cuts (Sect. 4.4) are used to get rid of big-Ms in the MP. The SP is a simple LP that checks feasibility. Strengthened nogood cuts are used based on minimal infeasible subsystems (Sect. 3.2.4).
- **Managing cranes in a container port** [49, 76, 200, 236, 237, 250]. In [76], the problem is to manage interfering and non-interfering cranes. It is assumed that gantry cranes are arranged linearly in groups so that only those within a group can interfere with each other. The objective is to minimize makespan. An enumerative branch-and-check procedure that uses only feasibility cuts (Sect. 4.3) is applied. The MP is an MILP that assigns jobs to cranes. The SP is a structured MILP, solved by a polynomial-time algorithm, that decouples by groups and checks whether there is a non-interfering schedule that improves makespan. • In [49], quay cranes are assigned tasks and scheduled. The MP decides for each ship the service start time, the number of quay cranes assigned to it, and the time of service completion. The SP checks whether cranes must cross and generates simple feasibility cuts (Sect. 3.2). • In [200], the problem is the dispatching of containers and conflict-free routing of yard cranes. The MP relaxes crane interferences and evaluates in what order, and by which crane, transportation requests are carried out to minimize makespan. The SP schedules crane movements. Simple optimality cuts (Sect. 3.2) are used. • In [236], the MP minimizes makespan by giving each crane a work zone consisting of consecutive ship bays, but it may also assign some cranes to other work zones, which can create conflicts between cranes. The SP minimizes makespan, given the MP solution, while taking into account conflicts with a cross-assigned crane. The SP decouples into groups of crane work zones, so that there can be no interference between cranes belonging to different groups. Simple nogood optimality cuts enforce a bound based on the current SP solution unless a work zone boundary is changed, or if no boundary is changed, some crane that is currently cross-assigned is no longer cross-assigned or vice-versa. • In [237], the model considers vessel stability as a ship is unloaded by multiple cranes at once, so that one side of ship is not heavier than the other. The MP assigns cranes to ships (bays), and the SP schedules unloading to preserve stability. Simple monotone feasibility and optimality cuts (Sect. 3.2.3) are used. • In [250], the MP assigns each ship a certain number of quay cranes that work in parallel. In so doing, it determines the value $\bar{x}_{i\ell t}$ of binary variables $x_{i\ell t}$ that indicate whether task i is assigned to ℓ cranes starting at time t. The objective is minimize the sum of weighted completion times of tasks, where the processing time for a task by ℓ machines is $p_{i\ell}$. The SP assigns each task to specific cranes and schedules tasks. A strengthened feasibility cut (Sect. 3.2.4)

$$\sum_{(i,\ell,t)\in S} x_{i\ell t} \leq |S| - 1$$

is used, where S is heuristically obtained by starting with $\{(i, \ell, t) \mid \bar{x}_{i\ell t} = 1\}$ and flipping to 0 the $\bar{x}_{i\ell t}$'s for which $p_{i\ell}$ is largest, until feasibility is achieved.

- **Berth allocation and ship loader scheduling** [45, 209]. In [45], the problem is set in a coal export terminal. The ships are loaded using a symmetrical berth layout-based rotary loading mode (SBLRLM) subject to deballasting restrictions. The MP uses MILP to determine the berthing position of each ship while ignoring berth capacity. The SP decouples by ship loader and uses CP to decide the sequence in which the loader handles loading tasks. An infeasible SP generates a monotone feasibility cut requiring that at least one ship/berth assignment be removed. A feasible SP generates a monotone optimality cut. Both cuts are strengthened by heuristically removing assignments in a method similar to Algorithm 1 in Sect. 3.2.4. A relaxation of the SP is included in the MP (Sect. 4.6). • In [209], time-varying water depth is taken into account. The MP assigns vessels to berths. The SP uses CP to sequence stevedoring tasks for the vessels assiged to each berth. Simple monotone feasibility and optimality cuts (Sect. 3.2.3) are generated.
- **Managing container stacking and relocation in a seaport** [50]. The problem is to manage the operation of automated cranes that stack and retrieve containers, so as to minimize the amount of container relocation when retrieving containers, as well as to schedule relocation when necessary. Each block of stored containers has two stacking cranes above it. Relocation consists of moving containers that are on top of containers to be loaded. The MP in a branch-and-check algorithm (Sect. 4.2) assigns jobs to cranes and sequences them so as to minimize makespan. The SP takes into account interference between cranes and precedence constraints of jobs assigned to the same crane. Simple monotone optimality cuts (Sect. 3.2.3) are used.

5.4.4   End User Delivery

- **Package delivery with multiple drones** [35, 147, 208]. In [35, 147], there is a single truck, carrying several drones. In [35], the MP assigns and sequences customers to be reached by the truck. The SP assigns and sequences customers to be reached by drones in multiple trips. Both the MP and SP are MILP problems. Simple nogood optimality cuts are used (Sect. 3.2.1). A SP relaxation is included in the MP by introducing some variables and constraints from the SP (Sect. 4.6.1). • In [147], the MP again assigns the truck to a route that covers a subset of demand points, and the SP covers the remaining demand points with drones of various types. If the SP is infeasible due to insufficient battery range, a simple feasibility cut requires a change in the set of demand points through which the truck passes. Otherwise, a simple optimality cut is used. The MP includes a SP relaxation by importing some SP constraints without integrality restrictions on the variables (Sect. 4.6.1). • In [208], there are multiple trucks and drones, and energy consumption by drones is constrained. Robust optimization is used to account for variable energy consumption due to weather. The MP routes trucks and drones. It determines y and $\boldsymbol{\pi}$, where $y_{iwj}^k = 1$ if truck k releases a drone to customer w from node i to j, and $\pi_i^{kh} = 1$ if a drone at node i is released in time interval $[t_{h-1}, t_h]$. The SP checks energy consumption constraints for drones using a flow model. If a drone that flies from i to j during period $[\bar{h}, \bar{q}]$ exceeds the energy limit, there is a feasibility cut

$$\pi_i^{k\bar{h}} + y_{iwj}^k + \pi_j^{k\bar{q}} \leqslant 2$$

If $[0, T]$ is the time horizon, the cut can be strengthened to

$$\sum_{h=0}^{\bar{h}} \pi_i^{kh} + y_{iwj}^k + \sum_{q=\bar{q}}^{T} \pi_j^{kq} \leqslant 2$$

- **Direct deliveries from depot to customers** [77]. Various kinds of trucks are used to make depot-to-customer round trips, one per delivery. Sometimes two or more round trips can be combined in a single tour ("milk run") subject to tight time windows. The objective is to minimize the maximum delay between release time and delivery, using branch and check (Sect. 4.2) with strengthened feasibility cuts. The MP assigns trips and milk runs (sets of trips) to trucks. The SP is relaxed in the MP by excluding milk runs that clearly violate time windows (Sect. 4.6.2). The SP decouples by truck. Dynamic programming is used to find all infeasible permutations of trips assigned to a given truck (if any), and corresponding feasibility cuts are added, with dominated cuts removed.
- **Consolidating items ordered online** [266]. An online order may require sourcing items (stock keeping units or SKUs) from multiple warehouses. The objective is to consolidate these items for shipment to the customer so as to minimize the total number of packages shipped. The MP solves a multicommodity flow problem in which the SKUs are commodities. The SKUs flow over transshipment arcs between warehouses (to consolidate items) and from warehouses to customers. The SP assembles SKUs moving from a given warehouse to a given customer into packages, subject to weight and compatibility constraints. One of the MP variables is x_{sjk} which is 1 if SKU s is shipped from node j to node k in the transport network, where a node represents a warehouse or custromer. We let z_{jk} be the number of packages shipped from j to customer k. The SP decouples into a packing problem for each warehouse j and customer k and minimizes the number of packages. Two cuts are defined. A monotone optimality cut says

$$z_{jk} \geqslant v^*(\bar{x}) - \sum_{s \in \bar{S}_{jk}} (1 - x_{sjk})$$

where $v^*(\bar{x})$ is the optimal value of the SP, and $\bar{S}_{jk}$ is set of SKUs sent from j to k in the MP solution. There is also an analytical optimality cut

$$z_{jk} \geqslant v^*(\bar{x}) - \sum_{s,j'|s\in\bar{S},}(1 - x_{sj'j}) - \sum_{s,j'|s\in\bar{S}}x_{sjj'}$$

where first sum is over j' such that arc (j', j) is in the warehouse-to-warehouse network and SKU s originates at j', and second sum is over j' such that arc (j, j') is in the warehouse-to-warehouse network and SKU s originates at j. The cut is based on the fact that if no SKUs coming into warehouse j are removed or forwarded to another warehouse, then the number of packages sent from from warehouse j to customer k does not decrease.

- **Crowdshipping** [33, 207]. Crowdshipping uses online apps to connect package delivery tasks with private cars or local courier services. In [33], packages are assigned to Amazon warehouse employees to deliver on the way home from work, if they volunteer. The MP makes the assignments, and the SP finds a routing that neither is too long (in time) nor pays too little per hour. The SP solved as a traveling salesman problem. If the SP solution is infeasible, a simple nogood feasibility cut (Sect. 3.2.1) removes at least one package from the assignment if the route is too long, and removes a package or adds a package if the assignment pays too little. A SP relaxation is added to the MP by using a warm start that pre-generates cuts for all possible assignments of limited size (namely 4) that are infeasible for a given employee. • In [207], both company trucks and private cars are used, and travel times are uncertain. Private cars are used only when a company truck cannot deliver on time. The MP assigns customer-to-customer arcs to trucks and private cars, and the SP obtains a schedule within time constraints. Monotone optimality cuts are used.

5.5   Computing and Telecommunications

5.5.1   Processing Task Assignment and Scheduling

- **Mapping task graphs to multicore architectures** [78–81]. The problem in [78, 79] is to map tasks in a directed acyclic graph (DAG) representing precedences to heterogeneous platforms. The MP assigns tasks to processors and a start time to each task, respecting precedences in the DAG. The SP begins with start times from the MP and checks whether tasks can be sequenced on each processor to observe nonoverlap. The SP variables are implicit (paermutation of tasks on each processor). Simple feasibility cuts (Sect. 3.2.3) are generated when the SP is infeasible. Analytic cuts are obtained by augmenting the DAG with arcs corresponding to precedences between unrelated tasks on each processor, and labeling each node with a start time. Then a cut is formed from tasks on a critical path in the resulting DAG. In addition, analytic knapsack cuts are generated, based on a time window relaxation when a processor's schedule is infeasible. • In [80], the MP contains a time window relaxation of the SP (Sect. 4.6.2), and strengthened feasibility

cuts are obtained heuristically using a variation of the heuristic methods in Sect. 3.2.4.
• In [81], a third, intermediate stage is added the the LBBD procedure (Sect. 4.7) that
checks SP feasibility subject to a subset of SP constraints. The final stage is the full SP.

- **Assigning and scheduling tasks in priority order** [40]. The MP uses CP to assign
  computational tasks to processors. The SP decouples by processor and schedules tasks
  on each processor in priority order until an infeasibility is found. A strengthened monotone feasibility cut is obtained using the QuickXplain algorithm without binary search
  (Algorithm 5 in Sect. 3.2.4). The cuts are imposed in the MP as not-all-equal constraints
  rather than inequalities.

- **Assigning and scheduling computational and memory tasks** [21, 22]. Three-stage
  LBBD (Sect. 4.7) is used. The SP is split into MP2 and SP2. The master problem MP of
  the primary LBBD algorithm assigns tasks to processors. Its objective function is heuristic
  and designed to spread tasks across processors. The primary LBBD algorithm is enumerative (Sect. 4.3) and terminates when MP1 becomes infeasible. Cuts for the MP are
  obtained from a subproblem that is solved by a secondary LBBD algorithm in which master problem MP2 assigns memory tasks to memory devices. MP2 decides whether a task's
  memory requirement is met locally (in the assigned processor) or remotely, and similarly
  for communication between processors required by the task. Strengthened monotone feasibility cuts (Sect. 3.2.1) in MP2 require that at least one memory requirement currently
  met locally is remote in future solutions. They are strengthened using heuristic binary
  search (Algorithm 4 in Sect. 3.2.3). Symmetry cuts are generated (Sect. 3.7).

- **Offloading tasks to multi-access edge computing** [6, 162]. In [6], the aim is to offload
  computational tasks from Internet of Things devices to applications hosted by edge
  servers. The MP assigns tasks to applications. The SP decouples by application and
  maximizes, for each application, the number of tasks feasibly scheduled on it. If one or
  more task remains unscheduled on a given application, a feasibility cut similar to (3.30)
  in Sect. 3.4.5 requires that assigning any one of the unscheduled tasks must result in
  the rejection of at least one of the scheduled tasks. • In [162], computational tasks are
  offloaded from smart toys to edge computing. Strengthened monotone cuts similar to
  the critical path cuts of [79] are used. See the above entry on mapping task graphs to
  multicore architectures.

- **Task allocation to cloudlets in mobile edge computing** [134]. The problem is formulated as an MILP to be solved by LBBD. The MP purchases computing resources and
  allocates tasks to cloudlets to maximize profit. The SP schedules tasks on cloudlets. The
  cuts appear to be simple feasibility cuts (Sect. 3.2) that require, for any cloudlet that
  cannot be scheduled, that resources be added or tasks removed. The problem is similar
  to that in [6].

- **Assigning families of tasks to processors viewed as 2-dimensional knapsacks** [173].
  The processors are constrained in 2 dimensions, such as CPU and RAM. Computing
  tasks come in families, and a task can be selected for processing only if its entire family
  are selected. Family members need not all be assigned to the same processor, but there is

a penalty for splitting a family. The MP decides which items will be selected for inclusion in some knapsack subject to aggregated knapsack capacities. The SP decides into which knapsack each task is placed. Simple monotone feasibility cuts (Sect. 3.2.3) are used.

- **Scheduling an avionics system** [149, 264]. Avionics refers to the electronic system on an aircraft. It consists of processors that perform computational tasks and a messaging system that allows the processors to communicate with each other. The messaging and task processing are often scheduled offline due to the complexity of the problem and the safety hazard of delayed communication. In [149], the schedule is cyclic, meaning that it repeats periodically. LBBD is applied to find a feasible schedule. The MP assigns messages to time slots, and the SP sequences tasks on each processor, subject to complex precedence constraints involving the tasks and the messages. Strengthened feasibility cuts are obtained by the DFBS method (Algorithm 3 in Sect. 3.2.4). Since the SP is not monotone, the cuts are strengthened by repeatedly solving a problem-specific relaxation of the SP that is designed so that if only some of the MP variables are fixed in the SP, an infeasible relaxed SP implies that the original SP is infeasible when any values are assigned to the unfixed variables. Also, a restriction of the SP is designed so that if the SP has a feasible solution when only some MP variables are fixed, that solution is feasible in the original SP. This provides a heuristic for finding feasible solutions. • In [264], the model represents a time-triggered protocol and minimizes bus bandwidth utilization. The MP is solved by a SATM (SAT modulo theory) solver, and the SP by geometric programming. Cuts are not explicitly described.

- **Information flow for autonomous cooperative autonomous driving** [232]. The problem is the optimal management of information flow to and from a fleet of autonomous vehicles under cooperative control, with the aim of supplying every vehicle with current data. The model is in discrete time. Each data stream is an uninterrupted series of communication episodes with the vehicles, where each episode has three stages: upload, computation, and download. The MP assigns radio and computational resources to each episode, and the SP schedules data transfers. Binary indicators are defined for upload, computation, and download processes. Each indicator is 1 if the resources assigned to a process are less than or equal to the resources assigned in the current MP solution. An analytical optimality cut is generated to require that the AoI (assessment of information efficiency) must be at least as great as in the SP subproblem solution if the indicators are equal to 1 for all episodes in the process. The computation of the indicators is complex and based on the precise structure of the problem.

5.5.2 Telecommunication Network Design

- **Local area network design** [99, 101, 196]. In [99, 101], the objective is a "green" design that minimizes power consumption. The MP selects wireless tower locations and assigns power levels, and the SP assigns customers to towers. The SP does not decouple. Feasibility cuts (Sect. 3.2) require that at least one tower that is not used must be put into service, or at least one tower that is in use must be given a higher power level, which allows more customers to connect to it. • In [101], the model is extended to problems that consider cost of assigning customers to a wireless tower, and it is solved by branch and check (Sect. 4.2). • In [196], a network design problem is solved by a 3-level decomposition (Sect. 4.7). An overall logic-based decomposition is used in which the MP is solved by MILP, and the SP generates simple feasibility and optimality cuts (Sect. 3.2). The SP is itself solved by classical Benders decomposition.

- **Three-colorable unit disk covers** [227]. A 3-colorable subcollection of unit disks must cover a set of points. The disks represent cell phone antennas, the points represent customers, and 3-colorability represents a noninterference requirement. The MP solves a set covering problem, and the SP checks colorability. The conflict graph representing the coloring SP is decomposed into biconnected components, which are collapsed to a point. This results in a block-cut tree. If the conflict graph is uncolorable, then some biconnected component is uncolorable. This is because a coloring of each component can be extended to a full covering by traversing the tree and permuting the colors of each component as necessary. Now the uncolorable components provide strong analytical feasibility cuts.

- **Locating regenerators in an optical network** [165]. The problem is to place a minimum number of regenerators in an optical network to ensure connectivity when any one arc is removed. Nodes can communicate along a generator-free path only if the path is not too long. The MP in a branch-and-check procedure (Sect. 4.2) locates regenerators at nodes by solving a set covering problem that relaxes the original problem. The SP checks whether there is a flow between every pair of nodes in a multicommodity model. This makes the SP an LP problem, and combinatorial Benders cuts are used (Sect. 4.4). An MILP is solved on the side to find a minimal infeasible constraint set. To make this practical, a smaller SP is obtained by reducing the number of fault scenarios that must be considered.

- **Edge partitioning for SONET rings** [239]. The problem is to partition the edges of a graph by defining subgraphs such that each edge appears in exactly one subgraph, and the endpoints of each edge appear in the same subgraph. Edges have weights, which are uncertain, and the weight of a subgraph is the sum of its edge weights. The edges represent communication channels between customers, and the edge weights are traffic volumes. The traffic in each subgraph is carried by a SONET ring. An upper bound UB on subgraph weights is imposed on every scenario. The paper uses a modification of the integer L-shaped method that is a special case of stochastic LBBD (Sect. 4.5). The MP determines binary variables v_{ijk}, where $v_{ijk} = 1$ when node i and/or j belongs to subgraph k. The SP decouples by scenario and determines the edge partition in each scenario subject to the bound UB. If the SP is infeasible, it is observed that the following monotone feasibility cut is valid:

$$\sum_{ijk|\bar{v}_{ijk}=0} v_{ijk} \geqslant 1$$

where $\bar{v}$ is the solution value of v in the MP. If a positive lower bound, namely $\text{LB}^{\bar{q}}$, on the maximum summed weight is obtained for some scenario $\bar{q}$, the cut can be strengthened to form the analytical cut

$$\sum_{ijk|\bar{v}_{ijk}=0} \min\{w_{ij}^{\bar{q}}, \text{LB}^{\bar{q}}\} v_{ijk} \geqslant \text{LB}^{\bar{q}}$$

where $w_{ij}^{\bar{q}}$ is the weight of arc (i, j) in scenario $\bar{q}$.

- **Reliable Network Design** [31]. Branch and check (Sect. 4.2) is applied to a stochastic network design problem for survivability, subject to constraints on path length (i.e., the number of "hops"). The MP fixes which arcs are used (with capacities). The SP is an IP problem that checks flow feasibility and decouples into one problem for each commodity. Classical Benders cuts are used, based on an LP relaxation of the SP. Simple nogood feasibility cuts (Sect. 3.2.1) are also defined for the SP. They require that one of the unused possible arcs be used in the network. The nogood cuts are found to be unnecessary for the instances tested when classical cuts are used.

- **Network upgrade** [65]. This "network migration" problem asks how to upgrade the arcs of a network by installing equipment at the endpoints of the arcs. It generates nogood cuts for general integer variables, but not as described in Sect. 3.2.1. The objective is to assign worker shifts to the endpoints of arcs so as to upgrade the arcs while minimizing cost. Several arcs may connect two endpoints i, i'. The MP decides how many shifts $x_{ii't}$ to assign to each pair i, i' in each period t. The SP decouples by period and decides *which* shifts to assign to each pair i, i'. Classical cuts are obtained from a solution by column generation of the LP relaxation of the SP, where the columns correspond to shifts. Logic-based cuts are obtained from a CP solution of the SP. If the SP is infeasible for a MP solution $\bar{x}$, monotone feasibility cuts (Sect. 3.2.3) require that at least one of the general integer variables $x_{ii't}$ that is positive be reduced by 1. The following cut is generated for each period t in which the SP is infeasible and each pair i, i':

$$x_{ii't} - M\delta_{ii't} \leqslant \bar{x}_{ii't} - 1$$

where M is a large number. The cut is enforced when 0–1 variable $\delta_{ii't} = 0$. To require that at least one $\delta_{ii't}$ must be 0, the following cut is added for each infeasible period t:

$$\sum_{i,i'|\bar{x}_{ii't}>0} (1 - \delta_{ii't}) \geqslant 1$$

Monotone optimality cuts (Sect. 3.2.3) of the following form are also generated for each period t:

$$z \geqslant v^*(\bar{x})\left(1 - \sum_{i,i'|\bar{x}_{ii't}>0}(1 - \delta_{ii't})\right)$$

where $v^*(\bar{x})$ is the optimal value of the SP.

5.6 Medical Applications

5.6.1 Scheduling and Staff Assignment

- **Clinical outpatient scheduling** [46, 218]. In [46], the focus is on chronic outpatients with comorbidities. The problem is to schedule patients with multiple conditions into a clinic. Each day is scheduled separately, using a rolling horizon. Each patient has a packet of services to be delivered on the same day. The MP uses answer set programming to assign days to patients. The SP schedules patients for each day, using a parallel machine scheduling model, and generates what are apparently simple nogood feasibility cuts (Sect. 3.2.3). • In [218], outpatient appointments are scheduled, where each patient is assigned a package consisting of a doctor, room, equipment, etc. The problem is solved by three-level LBBD (Sect. 4.7). The MP of the first level uses MILP to assign days to patients, and a feasibility SP assigns packages and schedules. The MP contains a relaxation of the SP based on energetic reasoning (Sect. 4.6.2). The cuts are simple monotone feasibility cuts (Sect. 3.2.3). The SP is partitioned into a MP that uses MILP to assign packages and a SP that uses MILP to assign schedules. The MP is again strengthened by energetic reasoning. The cuts are not described.

- **Scheduling of operating rooms with predicted surgery duration** [195, 220–224]. The problem is to schedule patients and medical staff into operating rooms (ORs). In [222], the objective is to minimize costs associated with opening surgical suites (which may contain several ORs) and patient waiting time. The MP decides which ORs to open on each day, and which which hospital to assign each patient on each day. The SP assigns patients to ORs and schedules surgeries. The SP decouples by hospital, surgical suite, and day. The MP is solved by MILP with symmetry breaking (due to interchangeable ORs). Binary variable $x_{hdp} = 1$ if patient p is assigned to some OR in hospital h on day d; y_{hd} is the number of ORs opened in hospital h on day d. The simple monotone feasibility cut

$$\sum_{p\in P_{hd}}(1 - x_{hdp}) \geqslant 1$$

and the simple monotone optimality cut

$$y_{hd} \geqslant \bar{y}_{hd} - \sum_{p\in P_{hd}}(1 - x_{hdp})$$

are used, where P_{hd} is the set of patients assigned to hospital h on day d, and $\bar{x}$, $\bar{y}$ are the solution values obtained in the MP. Symmetry cuts (Sect. 3.7) are also generated: when certain patients cannot be scheduled on a certain hospital-day, they cannot be scheduled on other hospital-days with equal or greater available OR time committed. Four implementation schemes are tested. A branch-and-check (Sect. 4.2) version of this approach is used in [220] to solve a problem with multiple levels of surgeon qualifications. Slightly different models appear in [220, 221]. Feasibility cuts in [220] require that fewer patients, more rooms, or more surgeons be assigned to a hospital-day. • In [195], the objective is to minimize overtime costs incurred by Toronto General Hospital. Surgery durations are estimated on the basis of an empirical microanalysis of the various activities that precede and comprise the surgery. Several MP/SP decompositions are tried, confirming the conventional wisdom that the computational load should be roughly equalized between the MP and SP. This results in a decomposition in which the MP assigns patients, anesthetists, surgeons, and ORs to days, and surgeons to patients. The SP assigns anesthetists, surgeons and ORs to patients, and sequences and schedules the surgeries. When staff cannot be scheduled on a given day, a disaggregated feasibility cut (Sect. 3.2) removes a patient assignment to the day or increases the staff allocation to the day. Similar disaggregated optimality cuts are generated when the staff schedule is feasible. • In [224], standard LBBD and branch and check are compared with a branch-price-and-cut method.

- **Scheduling of operating rooms with unpredictable surgery duration** [109, 146]. In [109], stochastic LBBD (Sect. 4.5) is used to schedule surgeries in multiple hospitals on multiple days, where the duration of surgeries is uncertain. The MP assigns patents to hospitals, operating rooms (ORs) in those hospitals, and days on which the surgery is to be performed. Some surgeries may be canceled, which means they are delayed to the next planning period. The SP decouples by hospital, OR, day, and probabilistic scenario. The MP minimizes various costs, including surgery cancellation costs. If surgery must be canceled for some patients in a scenario, the SP decides which patients to cancel to minimize cost. An analytic optimality cut imposes a lower bound equal to the minimum cancellation cost in the subproblem minus the cancellation cost for each patient removed from the assignments to that OR on that day (see Sect. 3.3). In addition, classical Benders cuts are generated from a decision diagram that represents a relaxation of the SP. • In [146], the problem is solved over a rolling horizon, again with stochastic surgery durations. The LBBD procedure is part of a heuristic. The MP assigns patients to "blocks," and the SP sequences surgeries in each block while assigning surgeons. Slack is included in the schedule to accommodate the arrival of emergency patients. If a surgeon must work on two patients in the same block on the same day, a simple monotone feasibility cut (Sect. 3.2.3) requires that one of them be deassigned from that block.

- **Hospital therapist scheduling** [141]. Branch and check (Sect. 4.2) is used in the context of column generation to solve an MILP model that assigns therapists to patients and therapy locations. The MP determines therapist assignments, room assignments, and therapy session start times subject to time windows. The SP checks precedence constraints and location capacity, where the latter depends on the timing of sessions. Simple monotone feasibility cuts require that at least one of the arcs used in the therapist tours be dropped. Since the problem is formulated for column generation, the cuts have the form

$$\sum_{r\in\Omega} \rho_r \lambda_r \leqslant |A| - 1$$

where r indexes columns, Ω is the current set of columns, ρ_r is the number of arcs encoded in column r, $\lambda_r = 1$ if column r is used, and A is the set of arcs in the therapist tours. This cut is valid, however, only if there is no way to adjust the start times so that the capacity and precedence constraints can be satisfied without changing the tours. An MILP side problem is solved to check for validity, although this step is found to have little effect in practice.

- **Assignment and scheduling of home healthcare aides** [108, 117, 118, 191]. In [117, 118], a home hospice care assignment and scheduling problem is solved, optionally on a rolling basis in which existing patients retain the same caregiver but may be rescheduled subject to constraints. Caregiver qualifications are matched to patient needs, and work schedule constraints are observed. Travel times between patients are taken into account. The objective is to maximize the number of patents served with a given staff of aides. Heuristically strengthened monotone optimality cuts (Sect. 3.2.4) are used, along with time window and other SP relaxations (Sect. 4.6) in the MP. • In [108], the existing patient schedules are fixed, which reduces the complexity of the problem and allows visit patterns to be enumerated in advance. The MP solves a covering problem that assigns patterns to patients. The patterns consist of patient, provider, set of visit days, and visit time. The SP is solved in 2 stages. The first stage considers only travel time and time window constraints. If the SP is infeasible, a feasibility cut is generated. Otherwise, the SP checks if work rules are satisfied (work time, overtime, etc.) and generates a feasibility cut if not. • In [191], the objective is to minimize caregiver cost, where potential visit days for patients are pre-specified. Branch and check first solves a deterministic problem and then finds a robust solution. The MP decides which caregiver serves which patients on which days, and the SP schedules the visits. Simple monotone optimality cuts are generated for each caregiver. A cut for a given caregiver is duplicated for caregivers with equal or higher qualifications (Sect. 3.7). The SP is solved approximately if it runs too long, which case its upper bound is retained, and the cut is pooled for later use. After branch and check terminates, the search branches on whether each pooled cut is to be enforced, solving SP to optimality. Pooled cuts are removed when they are no longer effective given current upper bound. A robust version of the problem is solved by including constraints that allow a restricted number of parameters to take their worst-case values.

5.6.2 Therapeutics

- **Finding repairable solutions for kidney exchange** [55]. The kidney exchange problem addressed here is the construction of exchange cycles in which each donor provides a suitable kidney for a relative of the next donor in the cycle. The goal is to find "super solutions," which are solutions in which any disruption causing a change in a bounded number of variables can be countered by changing a bounded number of other variables. The MP is the original optimization problem, and the SP is the repair problem. In the MP, binary variable $x_i = 1$ if cycle i is used in the solution. Assuming that one cycle is going to be broken in the solution $\bar{x}$ of the MP, the SP decouples into a repair problem for each possible disruption. In each repair problem, we determine the smallest number of $\bar{x}_i$s currently equal to 0 that must must be flipped to 1 to restore lost transplants, while flipping at most b variables. Let this min number be β^*. An analytic feasibility cut is generated to require that for a given x_k currently equal to 1, if cycle k fails ($x_k = 0$) we must change at least $b - \beta^*$ of the x_is in the MP solution, because we allow repairs that add no more than b new cycles. So the cut is

$$\sum_{i:\bar{x}_i=0} x_i + \sum_{i:\bar{x}_i=1}(1 - x_i) \geq (b - \beta^*)(1 - x_k)$$

- **DNA sequence alignment** [130]. The task is to align multiple DNA subsequences. The sequencing problem for each pair of subsequences is formulated as a multivalued decision diagram (MDD). The multiple sequencing problem is then formulated as an MILP that links the MDDs. The MILP problem is solved by LBBD in which feasibility cuts are generated when triples of subsequences fail to align. Alignment of all triples is sufficient to solve the multiple sequencing problem.
- **Radiation therapy control** [238]. Radiation is directed at cancer tissue by selecting the aperture through which the beam passes, and beam intensity. The MP decides which apertures are used, and the SP decides the intensity of the beam. The SP is an LP problem, and combinatorial Benders cuts (Sect. 4.4) are used.
- **Prostate cancer screening** [267]. A screening strategy is developed so as to balance the cost and inconvenience of testing with the possibility of detecting prostate cancer. Different strategies may be assigned to individuals with different profiles. The MP fixes $\bar{x}_{st} = 1$ when a biopsy is scheduled at period t in surveillance strategy s, and $\bar{y}_{ps} = 1$ when strategy s is assigned to patient type p. The SP computes the expected benefit, given $\bar{x}$ and $\bar{y}$. The SP yields an optimality cut that is based on the maximum loss created by restricting each patient type to its assigned testing strategy, as opposed to allowing every patient to have a different strategy (see Sect. 3.3).

5.6.3   Epidemics

- **Assignment and scheduling of equipment orders during an epidemic** [167]. The problem is to allocate orders for personal protective equipment to multiple manufacturers during an epidemic, so as minimize cost subject to parameterized bounds on makespan. The MP decides which orders to accept and assigns accepted orders to factories. The SP decouples by factory and schedules the orders to minimize makespan. Strengthened monotone optimality cuts are obtained by a deletion filter method (Sect. 3.2.4). Uncertainties in demand and production are accommodated by "predictive-reactive" scheduling, which creates a schedule based on predicted values and then updates the schedule dynamically as the situation changes.
- **Location of Covid-19 testing centers** [168, 263]. The objective is to locate testing sites to serve a population as conveniently as possible. In [168], a biobjective model similar to that in the previous entry is formulated, in which total travel distance is the primary objective and makespan the secondary objective. The MP is a p-median problem that locates testing sites to minimize total travel distance. The article states that cuts are generated as in [169] (see "Machine location, assignment, and scheduling" in Sect. 5.3.1). A pareto-optimal frontier is revealed as cuts are generated and the MP re-solved. A similar approach is described in [263].

5.7   Other Applications

5.7.1   Disaster Management

- **Robust disaster preparedness** [83]. The problem is to locate permanent and temporary facilities. Permanent facilities (PFs) house supplies, while temporary facilities (TFs) serve as distribution and refuge centers when disaster strikes. In each scenario, all constraints are enforced and demand points are assigned to closest TF. The MP locates PFs and TFs and assigns inventory to PFs. The SP has two parts. One part assigns demand points to the closest TF and decouples by scenario. It generates simple monotone optimality cuts (Sect. 3.2.3). The other part determines the supply flow distribution and decomposes by scenario and item type. If flow out of a TF p exceeds its capacity in scenario s, a cut forces one of two adjustments. (a) The TF p is shut down. (b) A closed TF p' that is closer than p to a demand point assigned p is opened.
- **Earthquake infrastructure risk management** [107]. A stochastic optimization model minimizes total expected cost of hardening infrastructure before an earthquake and repairing it after the earthquake, under numerous scenarios. The MP decides which bridges, utilities, etc., to harden against an earthquake before it occurs, and the SP minimizes

repair cost after it occurs. Simple monotone optimality cuts (Sect. 3.2.3) are used. The model is applied to the San Francisco Bay area.

- **Search and rescue after an earthquake** [261]. Rescue teams are needed to help trapped people in multiple affected sites, where the number of people saved depends on how soon a team arrives and how long it stays. The problem is to allocate and schedule teams to maximize the survival rate. In a branch-and-check (Sect. 4.2) procedure, the MP determines which sites to visit, which team visits each site, and their routes. The SP determines the service time at each site. Let $x_{ijt} = 1$ when team t travels directly from site i to j, and $y_{it} = 1$ when site i is visited by team t. The MP determines values $\bar{x}_{ijt}, \bar{y}_{it}$. It contains lexicographic ordering constraints to reduce symmetry. The SP decouples by team. The SP is initially assessed for feasibility by setting service times to zero and checking whether arrival times meet deadlines. Let $r_1, \ldots, r_h$ be the sites on an infeasible route for team t from the first visited site to the first at which deadline is missed. The analytic feasibility cut

$$\sum_{i=1}^{h-1} x_{r_i r_{i+1} t} \leqslant h - 2$$

is generated. If the SP is feasible, and analytical optimality cut is generated for each team t:

$$z_t \leqslant v_t^* + (v_t^* - \text{UB})\Big(\sum_{(i,j) \in v} x_{ijt} - \sum_i \bar{y}_{it} - 1 \Big)$$

where z_t represents the benefit created by team t, v_t^* is the current optimal value of the SP, and UB a known upper bound on the possible benefit. A method is given for tightening UB.

- **Fortifying service facilities** [263]. Customers are supplied by the nearest service facility in a network. These facilities can go down due to a natural disaster or an attack. If resources are available to fortify a limited number of facilities, the problem is to decide which to fortify so as to minimize the worst-case transportation cost of serving customers. The MP decides which facilities to fortify. The SP computes the worst-case cost imposed by an agent subject to the agent's resource limit. The SP objective function is highly nonlinear but supermodular, allowing the SP to be solved by a specialized cutting plane algorithm. The MP determines binary variables x_j, where $x_j = 1$ when facility j is fortified. If $\bar{x}$ is the MP solution, we let $J = \{j \mid \bar{x}_j = 1\}$. Then if $v^*(J)$ is the optimal SP value, the following optimality cut is valid:

$$z \geqslant v^*(J) + \sum_{j \notin J} \Delta_j^+(J) x_j + \sum_{j \in J} \Delta_j^-(J)(1 - x_j)$$

where $\Delta_j^+(J)$ is the largest possible (negative) change in worst-case cost that can result from fortifying a facility j not currently fortified, and $\Delta_j^-(J)$ is the largest possible (positive) change in worst-case cost that can result from not fortifying a facility currently fortified. Thus we have

$$\Delta_j^+(J) = \min_{S \supseteq J} \left\{ v^*(S \cup \{j\}) - v^*(S) \right\}, \quad \text{for } j \notin J$$

where J is the set of facilities, and similarly for $\Delta_j^-(J)$. Since computing $\Delta_j^+(J)$ requires exponentially many evaluations of v^*, it is replaced by a valid bound on $\Delta_j^+(J)$ that requires a linear number of evaluations, and similarly for $\Delta_j^-(J)$.

- **Electric grid restoration** [119, 228]. The task is to find an optimal reconfiguration of an electric grid to respond to outages. In [119], the MP uses MILP to determine which transmission lines are restored to service, and the SP is an LP that finds the optimal power output from each generator. If the grid topology is "revenue adequate," a classical Benders optimality cut (Sect. 2.4) is used. Otherwise, a simple nogood feasibility (Sect. 3.2) cut is generated. • In [228], the MP sets switches to restore power, whereupon the blackout area decouples into connected components, to be solved separately. The SP incorporates voltage and power flow constraints to find an optimal load pickup solution. The decision of which links to energize is the same through all time periods. An analytical feasibility cut considers what switching patterns would fail to restore connectivity and requires that some other switching pattern be used. A similar analytical optimality cut is obtained.
- **Wildfire suppression** [116]. The problem is related to the network interdiction problem described in [82]. The objective is to minimize the number of nodes in a network that are unprotected from fire within a given time horizon. The MP allocates resources to nodes, which slows the rate of fire spread on outgoing arcs by a known amount. The SP computes a shortest fire path to each node to which resources are allocated. The MP minimizes total node cost, where the cost of a node is 1 if the node is not protected. An analytic optimality cut says that the cost of a given node n that is unprotected in the SP solution is again 1 unless some unresourced node along fire path to n is resourced before fire reaches it. Another cut observes how many interdictions are necessary along the shortest fire path to n to protect n and requires that at least this many interdictions be used to reduce the node cost below 1. The cut should be checked carefully for validity.

5.7.2   Packing and Cutting

- **Strip packing** [62–64, 71, 182]. The object is to pack rectangles into a rectangular space, without rotation or overlap. There may be an objective function, such as minimizing the height of the containing rectangle. In [62], the MP in a branch-and-check procedure (Sect. 4.2) determines the x-coordinate of the left side of each rectangle to be packed. The SP is a feasibility problem that tries to find a feasible set of y-coordinates for the rectangles. Analytic feasibility cuts and a lifting method specialized to this type of problem are generated as described in Sect. 3.6.   • A similar method, without lifting, is used in [63] to solve the problem with unloading constraints. Optimality cuts strengthened

with one of the methods in [62], plus an additional lifting technique, are used in [64] to pack rectangles into a collection of larger rectangles. The SP is solved by CP in [71].
• In [182], three "guillotine cuts" are allowed. Analytic optimality cuts are obtained by introducing integer variables.

• **Packing orders into parcels for shipment** [92]. Items ordered online are packed into boxes for shipping so as to minimize wasted space. The MP assigns items to parcels, and the subproblem solves the 3D packing problem that results for each parcel. A specialized branch-and-check method (Sect. 4.2) is used with combinatorial Benders cuts (Sect. 4.4), except that infeasible solutions are repaired when possible by solving an auxiliary sub-problem.

• **Time-sensitive guillotine cutting of rectangular boards** [205]. A frequent problem in furniture manufacturing is to cut small rectangular items from large rectangular boards. The problem here takes in account earliness and tardiness costs, since the small items are needed at a particular time. The MP in a branch-and-check procedure (Sect. 4.2) assumes that the required items are dimensionless. The SP looks at the length and width of items and their pairwise compatibilities and relationship to shape of the boards. In the MP, $x_{ib} = 1$ when item i assigned to board b. The MP solved by both CP and MILP, but MILP is better if a warm start is used. Analytical feasibility cuts of the form

$$\sum_{i \in I_b \setminus \{j\}} x_{ib} + \frac{1}{|I^*|} \sum_{i \in I^*} x_{ib} \leqslant |I_b| - 1$$

are generated for each item $j \in I_b$, where I_b is the set of items currently assigned to board b, and I^* is set of items not assigned to board b that are equal to or larger than item j. So the cut excludes any assignment in which all items in I_b except j, and an item outside I_b at least as large as j, are assigned to board b.

5.7.3 Scheduling

• **Tournament scheduling** [211, 212, 252]. In [211], a triple round-robin tournament is scheduled, subject to complicated constraints, for the Danish soccer league. The tourna-ment is divided into two consecutive parts, the first a single round robin and the second a double round robin. The MP determines an assignment of patterns to teams in each part. The SP consists of two stages. The first stage checks a series of necessary conditions for combinatorial feasibility of part 1 assignments and generates feasibility cuts if they are failed, returning to the MP for each failure sequentially. If no cuts are generated, the second stage generates a schedule for the entire MP solution. Cuts include a variety of monotone feasibility and optimality cuts (Sect. 3.2.3). A similar approach is used in [212]. • In [252], games are scheduled in a round robin tournament so as to inject some fairness in the length of time gaps between games. The MP uses MILP to decide in which slots each team plays a game (home or away). The SP uses CP to decide whether a game is home or way and which team the assigned team plays. The SP does not decouple and generates three types of analytical feasibility cuts (Sect. 3.4) that are specific to this problem.

- **Baseball umpire scheduling** [248, 249]. The task is to assign umpires (referees) to baseball games in each time slot throughout the season, subject to complex fairness rules. A heuristic method that uses logic-based cuts is applied. The objective is to minimize cost, measured by the total distance traveled by umpires from one game to the next. A greedy heuristic finds, for each slot sequentially, a minimum-cost perfect matching between the set of umpires and the set T of games in that slot. Let G be the bipartite graph that represents the matching problem for the current slot. If there is no perfect matching in G, an analytic feasibility cut is generated for any set S of umpires (or games) whose adjacency neighborhood $N(S)$ in G has cardinality less than $|S|$; such a set S must exist due to Hall's theorem. The cut requires that at least one edge be added between S and $T \setminus N(S)$ by forcing a change of at least one assignment in previous slots that prevents G from containing edges between S and $T \setminus N(S)$.
- **Post-enrollment course timetabling** [39]. The MP uses CP to create a timetable. The SP checks whether rooms are available for the timetable. If not, a cut similar to a combinatorial Benders cut (Sect. 4.4) is generated, based on the QuickXplain procedure [142] (Algorithm 6). The cut takes the form of not-all-equal constraint in CP model rather than an inequality.
- **Multiphase course timetabling with cohort constraints** [85]. Each course consists of several phases or lessons, and each student enrolled in the course must work through all the lessons. The problem is to schedule the lessons with cohort constraints, meaning that certain students must attend certain lessons together as a cohort. The MP formulates the problem without cohort constraints, and the SP checks if the cohort constraints are satisfied. Branch and check (Sect. 4.2) is used with simple feasibility cuts (Sect. 3.2).
- **Robust call center scheduling** [58]. The aim is to find a robust project schedule (Sect. 4.5) for an IT services delivery center in which there are unpredictable delays, so as to minimize tardiness while informing the customer of a reasonable worst-case completion time. An empirically determined polyhedral uncertainty set is used. The problem is solved with three-level LBBD (Sect. 4.7) that generates heuristically strengthened monotone optimality cuts (Sect. 3.2.4).

5.7.4 Miscellaneous

- **Verifying logic circuits** [127]. This can be seen, in retrospect, as the first application of LBBD. The problem is to check whether a logic circuit is correctly designed by comparing its output with that of a circuit (or Boolean function) that is known to be correct. This, in turn, is accomplished by checking whether their outputs are the same for all possible inputs. The MP variables are the inputs, and the SP rapidly computes the output by tracing signals through the logic gates in the circuit. The output is 1 of the two circuits are equivalent, and the problem is to minimize the output. If the minimum is 0, the circuits are not equivalent. If the output is 1 for a given input, one or more subsets of the current inputs sufficient to yield the output are identified by tracing the signals back through the circuit. This yields a conflict clause (Sect. 3.8) that serves as an analytic feasibility cut.

- **Learning optimal decision trees** [32]. The problem is to design a decision tree in which the split at each node is defined by a separating hyperplane rather than value of a single feature. The MP defines the shape of the tree and how observations are routed through it. The linear SP seeks a separating hyperplane that captures the split defined by the MP at each node of the tree. If some minimal subset of observations cannot be split in this way ("shattered"), a nogood cut similar to a combinatorial Benders cut (Sect. 4.4) is introduced to force at least one of the observations to be routed to the other side.

- **Network interdiction** [82]. The task is to cut flow on a limited number of arcs so as to minimize maximum total flow over multiple periods subject to a cost bound. The problem is motivated by interdiction of drug traffic. The MP uses MILP to select arcs for interdictions, and the SP uses CP to schedule the interdictions. Simple nogood cuts are used in a branch-and-check framework (Sect. 4.2).

- **Military flow diversion** [160]. The problem is to destroy arcs so as to force all source-to-sink flow through a single predetermined "diversion arc" where the enemy may be attacked. There may be multiple sources and sinks. The MP is a branch-and-check procedure that decides whether there is a source-to-sink (s–t) path through the diversion arc (i, j) for each pair s, t. If not, then for some s, t, there is no path from s to i or from j to t; suppose the former. Then a minimum s–i cut is computed, and a feasibility cut written to exclude at least one arc in this cut from subsequent cuts obtained in the MP.

- **Placing sensors to detect water system contamination** [128]. In a water distribution network, certain nodes are vulnerable to contamination. A sensor placed downstream from such a node can detect contamination at that node. Let S_i be the set of sensors that are downstream from vulnerable node i. Let two vulnerable nodes i, i' belong to the same cluster if $S_i = S_{i'}$, meaning that sensors cannot distinguish which of the two is contaminated. Ideally, all clusters should have cardinality 1, but this is too expensive. The objective is to minimize the cardinality z of the largest cluster, subject to budget constraints. If z^* is optimal, a secondary objective is to minimize the number of clusters of size z^*, then minimize the number of size $z^* - 1$, and so forth. The MP places sensors at nodes so as to minimize z, subject to budget limitations and the condition that every vulnerable node have at least one downstream sensor. The subproblem determines the resulting clusters. Let UB be the smallest maximum cluster size obtained so far in the LBBD algorithm. Let $\delta_j = 1$ when the MP places a sensor at node j, let T_i be the set of nodes downstream from vulnerable node i, and let C_k be the kth cluster in the subproblem solution. Then if the size of cluster C_k is greater than UB, the following enumerative cut (Sect. 4.3) is imposed:

$$\sum_{j \in D_k} \delta_j \geqslant 1, \quad \text{where} \quad D_k = \bigcup_{i,i' \in C_k} T_i \ominus T_{i'}$$

and where $\ominus$ is symmetric difference. If cluster C_k has size UB, the following optimality cut is imposed:

$$z \geqslant \text{UB} - M \sum_{j \in D_k} \delta_j$$

where M is a large number. The cuts are based on the fact that if no additional sensors are placed in set D_k, there will be a cluster with cardinality at least $|C_k|$.

- **Energy policy analysis** [96]. A position paper that describes a framework for energy policy analysis. The MP selects allocation of energy sources. The SP runs a simulation to determine if some incentive structure will yield the desired production levels. The SP solution is based on regression analysis and is treated as an exact test of feasibility. The cuts not detailed.
- **Equilibrating electricity prices** [140]. LBBD is used to determine equilibrium electricity prices. The MP determines production from distributed energy resources. The SP decouples into community-level demands. Modeling details and cuts are not given; only numerical results.

5.8   Abstract Problem Classes

- **General 0–1 programming with subproblem decoupling** [126]. The intended application is to 0–1 programming problems in which the SP decouples. The decoupled components of the SP are solved by branch and bound. If a component is infeasible, the classical dual solution of the infeasible LP relaxation at each leaf node specifies a linear combination of constraints (i.e., a surrogate) that is violated by the variables fixed at that node. A logical conflict clause (Sect. 3.8) is derived from each surrogate, and the disjunction of the conflict clauses serves as a feasibility cut. If a component is feasible, dual solutions specify surrogates that prove the optimal value, and an optimality cut is similarly formed.
- **General optimal control** [30]. An MILP model for a general optimal control problem is solved by LBBD. The SP contains Boolean constraints (representing, for example, an automaton and design constraints). The SP is a feasibility problem solved by a SAT algorithm. Simple nogood feasibility cuts (Sect. 3.2.1) are used. The model is applied to a simplified motorcycle control problems with discrete states (gears, idle) and an automaton for state transitions.

- **Linear complementarity and quadratic programming** [14, 131, 132, 137]. In [132], the problem is to solve an LP problem with linear complementarity constraints. The problem is formulated as an MILP problem. Feasibility cuts in the form of logical clauses are derived from extreme points and rays of the dual of the LP relaxation of the MILP for the current values $\bar{x}$, $\bar{y}$ of 0–1 variables x and y in a complementarity constraint $xy = 0$. The cuts are heuristically strengthened by partitioning the literals of these clausal cuts into two subsets, each yielding a stronger cut ("sparsification"). The sparser cuts are checked for validity by whether the optimal optimal value of the LP relaxation of the original MILP augmented with the cut is greater than current upper bound. If so, the cut is judged valid because it cannot cut off the optimal value of the original problem. The strengthening procedure continues until no further valid sparse cuts are found. Rejected cuts are stored along with the corresponding relaxation bounds, in case they become admissible as the current UB is reduced. • The method is generalized in [14] to convex quadratic programs with complementarity constraints, and in [131] to nonconvex quadratic problems not known to be bounded. The method is further modified in [137] to resemble branch and check (Sect. 4.2) somewhat.
- **Automated planning** [69, 70, 145]. The automated planning problem is to specify a sequence of operations that achieve a desired goal while minimizing cost, where each use of an operator incurs a cost. A frequent application is to robot motion planning. In [69, 70], a form of LBBD similar to branch and check (4.2) is used to combine operator count heuristics with operator sequencing. The MP is an MILP model that heuristically computes bounds on the number of times each operator is applied. The solution of the MP is encoded in binary variables of the form $[Y_o \geqslant k_o]$, which are equal to 1 when the number Y_o of occurrences of operator o is at least k_o. The SP is a SAT formulation of the problem of sequencing operators to achieve the desired goal, subject to the bounds on counts obtained in the MP. If the SP is infeasible, a feasibility cut (known as a "landmark") is generated, having the form

$$\sum_{o \in L} [Y_o \geqslant k_o] \geqslant 1$$

  In [145], the SAT formulation of the SP is replaced with A* search.
- **Modular linear integer arithmetic** [144]. The task is to test the satisfiability of a set of inequalities $E_i \leqslant F_i$ in modular arithmetic with modulus m. The problem is solved by writing the constraints as $E_i + q_i m \leqslant F_i + r_i m$. The MP selects the quotients q_i, r_i. The SP uses IP to check for satisfiability. If the SP is unsatisfiable given quotients $\bar{q}_i, \bar{r}_i$, a strengthened feasibility cut is generated, consisting of a disjunction of $r_i - q_i > \bar{r}_i - \bar{q}_i$ over all constraints i in the unsatisfiable core obtained from the IP solver.
- **Minimum chordal completion** [23]. The MP minimizes the addition of new edges (chords) subject to Benders cuts. SP contains no new variables but applies an algorithm that identifies chordless cycles. An analytical feasibility cut is generated based on a theorem that completion a cycle of n vertices requires at least $n - 3$ new edges. So one might say this is LBBD despite being lazy constraint generation.

- **Minimum connected dominating set** [100]. Given a graph G, the objective is to find a minimum dominating set D of vertices (one such that every vertex in G is adjacent to a vertex in D) such that the subgraph of G induced by D is connected. The MP finds a minimum dominating set. The SP checks for connectedness. An analytical feasibility cut requires that just enough vertices be added to D to make D connected. This is determined by solving a Steiner tree problem on a graph obtained by shrinking each connected component of G to a point, and in which these connected components are the terminals of the Steiner tree. The number m of Steiner vertices in the minimum Steiner tree is the minimum number of vertices that must be added to the dominating set to achieve connectedness. The analytical cut is

$$\sum_{i\,|\,\bar{x}_i=0} x_i \geqslant m$$

where $x_i = 1$ indicates that vertex i is in the dominating set, and $\bar{x}$ is the solution of the current master problem.

- **Robust optimization with information discovery** [204]. The MP decides which information to observe, and the SP solves an optimization problem based on this information. Setting $w_i = 1$ allows one to observe uncertain element s_i. The problem can be stated $\min_{\boldsymbol{w}}\{\Phi(\boldsymbol{w})\}$, where

$$\Phi(\boldsymbol{w}) = \max_{\bar{s}} \left\{ \min_{\boldsymbol{y}} \left\{ \max_{s \in S(\boldsymbol{w},\bar{s})} (s^{\mathsf{T}} C \boldsymbol{w} + s^{\mathsf{T}} P \boldsymbol{y}) \right\} \right\}$$

and $S(\boldsymbol{w},\bar{s}) = \{s \mid s_i = \bar{s}_i \text{ for all } i \text{ with } w_i = 1\}$, where s ranges over scenarios. The vector $\bar{s}$ represents a scenario that materializes before decisions $\boldsymbol{y}$ are made. Thus, the innermost max is the worst-case cost over components s_i that were not observed ($w_i = 0$). The problem is decomposed by letting the MP be $\min_{\boldsymbol{w}}\{\Phi(\boldsymbol{w})\}$, and letting the SP be the problem of computing $\Phi(\boldsymbol{w})$. The cuts are simple optimality cuts (Sect. 3.2)

$$\Phi(\boldsymbol{w}) \geqslant \Phi(\bar{\boldsymbol{w}}) - \big(\Phi(\bar{\boldsymbol{w}}) - \underline{\Phi}\big)\phi(\bar{\boldsymbol{w}}, \boldsymbol{w})$$

where $\underline{\Phi}$ is a lower bound on Φ, and $\phi(\bar{\boldsymbol{w}}, \boldsymbol{w}) = \sum_{i\,|\,\bar{w}_i=1}(1 - w_i) + \sum_{i\,|\,\bar{w}_i=0} w_i$. If the information cost $s^{\mathsf{T}} C \boldsymbol{w}$ is certain, we have monotonicity (Sect. 3.2.3), and the cut is strengthened by letting $\phi(\bar{\boldsymbol{w}}, \boldsymbol{w}) = \sum_{i\,|\,\bar{w}_i=0} w_i$.

- **Piecewise linear regression** [256]. A continuous piecewise linear regression function is derived to model discrete data, allowing for outliers. The MP is an MILP model that decides in which segment each data point is located, and which points are selected as outliers. The SP is an LP problem that determines the location of break points and coefficients for each segment. The problem is solved by a standard application of combinatorial Benders cuts (Sect. 4.4).

- **Maximum satisfiability** [13, 66, 67]. The MAXSAT problem is to fix the values of binary variables so as to maximize the summed weights of true clauses in a given set of logical clauses. This can be viewed as a weighted hitting set problem. An artificial variable is associated with each clause indicating whether it is true or false. The objective is to max weighted sum of variables. In [13], the MP is an MILP model that maximizes this value subject to feasibility cuts. The SP conducts a "relaxation search," which is a branching tree that branches first on the artificial variables. When infeasibility is encountered, conflict clauses containing artificial variables are learned as in SAT solvers (Sect. 3.8) and added to MP as Benders cuts. Enhancements to the method are described in [66, 67].

References

1. Afsar HM (2022) Traveling salesperson problem with unique pricing and stochastic thresholds. Comput Ind Eng 2. https://doi.org/10.1016/j.cie.2022.108696
2. Afsar HM, Hnaien F (2020) Formulations and solution algorithms for dynamic assembly routing problem. Int J Prod Res 58:671–688. https://doi.org/10.1080/00207543.2019.1588481
3. Agra A, Christiansen M, Figueiredo R, Hvattum LM, Poss M, Requejo C (2013) The robust vehicle routing problem with time windows. Comput Oper Res 40:856–866. https://doi.org/10.1016/j.cor.2012.10.002
4. Agussurja L, Kumar A, Lau HC (2018) Resource-constrained scheduling for maritime traffic management. In: National conference on artificial intelligence (AAAI 2018). AAAI Press, pp 6086–6093. https://doi.org/10.1609/aaai.v32i1.12086
5. Alaei MRK, Soysal M, Elmi A, Banaitis A, Banaitiene N, Rostamzadeh R, Javanmard S (2021) A Benders algorithm of decomposition used for the parallel machine problem of robotic cell. Mathematics 9. https://doi.org/10.3390/math9151730
6. Alameddine H, Sharafeddine S, Sebbah S, Ayoubi S, Assi C (2019) Dynamic task offloading and scheduling for low-latency IoT services in multi-access edge computing. IEEE J Sel Areas Commun 37:668–682. https://doi.org/10.1109/JSAC.2019.2894306
7. Alvo M, Angulo G, Klapp MA (2021) An exact solution approach for an electric bus dispatch problem. Transp Res Part E 156. https://doi.org/10.1016/j.tre.2021.102528
8. Angulo G, Ahmed S, Dey SS (2016) Improving the integer L-shaped method. INFORMS J Comput 28:483–499. https://doi.org/10.1287/ijoc.2016.0695
9. Aron I, Hooker JN, Yunes TH (2004) SIMPL: a system for integrating optimization techniques. In Régin JC, Rueher M (eds) CPAIOR 2004 proceedings. Springer, pp 21–36. https://doi.org/10.1007/978-3-540-24664-0_2
10. Attar SF, Mohammadi M, Pasandideh S, Naderi B (2022) Formulation and exact algorithms for electric vehicle production routing problem. Expert Syst Appl 204. https://doi.org/10.1016/j.eswa.2022.117292
11. Avgerinos I, Mourtos I, Vatikiotis S, Zois G (2022) Scheduling unrelated machines with job splitting, setup resources and sequence dependency. Int J Prod Res. https://doi.org/10.1080/00207543.2022.2102948

J. Hooker, *Logic-Based Benders Decomposition*, Synthesis Lectures on Operations Research and Applications, https://doi.org/10.1007/978-3-031-45039-6

12. Avgerinos I, Mourtos I, Zois G (2021) Logic-based Benders decomposition for an inter-modal transportation problem. In: Stuckey P (ed) CPAIOR 2021 proceedings. Springer, pp 315–331. https://doi.org/10.1007/978-3-030-78230-6_20

13. Bacchus F, Davies J, Tsimpoukell M, Katsirelos G (2014) Relaxation search: a simple way of managing optional clauses. In: National conference on artificial intelligence (AAAI 2004), pp 835–841. https://doi.org/10.1609/aaai.v28i1.8849

14. Bai L, Mitchell J, Pang JS (2013) On convex quadratic programs with linear complementarity constraints. Comput Optim Appl 517–554. https://doi.org/10.1007/s10589-012-9497-4

15. Bajestani MA, Beck JC (2011) Scheduling an aircraft repair shop. In: Proceedings of the international conference on automated planning and scheduling (ICAPS). AAAI Press, pp 10–17. https://doi.org/10.1609/icaps.v21i1.13450

16. Bajestani MA, Beck JC (2013) Scheduling a dynamic aircraft repair shop with limited repair resources. J Artif Intell Res 47:35–70. https://doi.org/10.1613/jair.3902

17. Bajestani MA, Beck JC (2015) A two-stage coupled algorithm for an integrated planning and flowshop scheduling problem with deteriorating machines. J Sched 18:471–486. https://doi.org/10.1007/s10951-015-0416-2

18. Baptiste P, Pape CL, Nuijten W (2001) Constraint-based scheduling: applying constraint programming to scheduling problems. Kluwer, Dordrecht

19. Beame P, Kautz H, Sabharwal A (2003) Understanding the power of clause learning. In: International joint conference on artificial intelligence (IJCAI 2003)

20. Benders JF (1962) Partitioning procedures for solving mixed-variables programming problems. Numer Math 4:238–252. https://doi.org/10.1007/BF01386316

21. Benini L, Lombardi M, Mantovani M, Milano M, Ruggiero M (2008) Multi-stage Benders decomposition for optimizing multicore architectures. In: Perron L, Trick MA (eds) CPAIOR 2008 proceedings. Lecture notes in computer science, vol 5015, pp 36–50. Springer (2008). https://doi.org/10.1007/978-3-540-68155-7_6

22. Benini L, Lombardi M, Milano M, Ruggiero M (2011) Optimal resource allocation and scheduling for the CELL BE platform. Ann Oper Res 184:51–77. https://doi.org/10.1007/s10479-010-0718-x

23. Bergman D, Raghunathan AU (2015) A Benders approach to the minimum chordal completion problem. In: Michel L (ed) CPAIOR proceedings. Lecture notes in computer science, vol 9075. Springer, pp 47–64. https://doi.org/10.1007/978-3-319-18008-3_4

24. Bertsimas D, Brown DB, Caramanis C (2011) Theory and applications of robust optimization. SIAM Rev 53:464–501. https://doi.org/10.1137/080734510

25. Bertsimas D, Den Hertog D (2022) Robust and adaptive optimization. Dynamic ideas, Waltham, MA

26. Birge JR, Louveau FV (2011) Introduction to stochastic programming. Springer, New York. https://doi.org/10.1007/978-1-4614-0237-4

27. Bockmayr A, Hooker JN (2005) Constraint programming. In: Aardal K, Nemhauser G, Weismantel R (eds) Discrete optimization, Handbooks in operations research and management science, vol 12. Elsevier, Amsterdam, pp 559–600

28. Bolusani S, Ralphs TK (2022) A framework for generalized Benders' decomposition and its application to multilevel optimization. Math Program 196:389–426. https://doi.org/10.1007/s10107-021-01763-7

29. Booth KEC, Tran TT, Beck JC (2016) Logic-based decomposition methods for the travelling purchaser problem. In: Quimper CG (ed) CPAIOR 2016 proceedings. Lecture notes in computer science, vol 9678. Springer, pp 55–64. https://doi.org/10.1007/978-3-319-33954-2_5

30. Borzabadi AH, Sadjadi ME (2009) Optimal control of hybrid systems by logic-based Benders decomposition. In: Giua A, Mahulea C, Silva M, Zaytoon J (eds) Analysis and design of hybrid systems, vol 3, pp 104–107. https://doi.org/10.3182/20090916-3-ES-3003.00019

31. Botton Q, Fortz B, Gouveia L, Poss M (2013) Benders decomposition for the hop-constrained survivable network design problem. INFORMS J Comput 25:13–26. https://doi.org/10.1287/ijoc.1110.0472

32. Boutilier J, Michini C, Zhou Z (2022) Shattering inequalities for learning optimal decision trees. In: Schaus P (ed) CPAIOR 2022 proceedings. Lecture notes in computer science, vol 13292. Springer, pp 74–90. https://doi.org/10.1007/978-3-031-08011-1_7

33. Boysen N, Emde S, Schwerdfeger S (2022) Crowdshipping by employees of distribution centers: optimization approaches for matching supply and demand. Eur J Oper Res 296:539–556. https://doi.org/10.1016/j.ejor.2021.04.002

34. Bruglieri M, Mancini S, Peruzzini R, Pisacane O (2021) The multi-period multi-trip container drayage problem with release and due dates. Comput Oper Res 125. https://doi.org/10.1016/j.cor.2020.105102

35. Bruni M, Khodaparasti S, Moshref-Javadi M (2022) A logic-based Benders decomposition method for the multi-trip traveling repairman problem with drones. Comput Oper Res 145. https://doi.org/10.1016/j.cor.2022.105845

36. Cadoli M, Patrizi F (2006) On the separability of subproblems in Benders decompositions. In: Beck JC, Smith BM (eds) Integration of AI and OR techniques in constraint programming for combinatorial optimization problems (CPAIOR 2006). Lecture notes in computer science, vol 3990. Springer, pp 74–88. https://doi.org/10.1007/11757375_8

37. Cadoli M, Patrizi F (2009) On the separability of subproblems in Benders decompositions. Ann Oper Res 171:27–43. https://doi.org/10.1007/s10479-008-0383-5

38. Çalık H, Fortz B (2019) A Benders decomposition method for locating stations in a one-way electric car sharing system under demand uncertainty. Transp Res Part B 125:121–150. https://doi.org/10.1016/j.trb.2019.05.004

39. Cambazard H, Hebrard E, O'Sullivan B, Papadopoulos A (2012) Local search and constraint programming for the post enrolment-based course timetabling problem. Ann Oper Res 194:111–135. https://doi.org/10.1007/s10479-010-0737-7

40. Cambazard H, Hladik PE, Déplanche AM, Jussien N, Trinquet Y (2004) Decomposition and learning for a hard real time task allocation problem. In: Wallace M (ed) Principles and practice of constraint programming (CP 2004). Lecture notes in computer science, vol 3258. Springer, pp 153–167

41. Cambazard H, Jussien N (2005) Identifying and exploiting problem structures using explanation-based costraint programming. In: Barták R, Milano M (eds) CPAIOR 2005 proceedings. Lecture notes in computer science, vol 3524. Springer, pp 94–109. https://doi.org/10.1007/11493853_9

42. Cambazard H, Jussien N (2005) Integrating Benders decomposition within constraint programming. In: van Beek P (ed) Principles and practice of constraint programming (CP 2005). Lecture notes in computer science, vol 3668. Springer, pp 752–756. https://doi.org/10.1007/11564751_58

43. Cambazard H, Jussien N (2006) Identifying and exploiting problem structures using explanation-based constraint programming. Constraints 11:295–313. https://doi.org/10.1007/s10601-006-9002-8

44. Cao JX, Lee DH, Chen JH, Shi Q (2010) The integrated yard truck and yard crane scheduling problem: Benders' decomposition-based methods. Transp Res Part E 46:344–353. https://doi.org/10.1016/j.tre.2009.08.012

45. Cao Z, Wang W, Jiang Y, Xu X, Xu Y, Guo Z (2022) Joint berth allocation and ship loader scheduling under the rotary loading mode in coal export terminals. Transp Res Part B 162:229–260. https://doi.org/10.1016/j.trb.2022.06.004

46. Cappanera P, Gavanelli M, Nonato M, Roma M (2022) A decomposition approach to the clinical pathway deployment for chronic outpatients with comorbidities. In: Amorosi L, Dell'Olmo P, Lari I (eds) Optimization in artificial intelligence and data sciences. AIRO Springer series, vol 8. Springer, pp 213–226. https://doi.org/10.1007/978-3-030-95380-5_19

47. Castellucci PB, Costa AM, Toledo F (2021) Network scheduling problem with cross-docking and loading constraints. Comput Oper Res 132. https://doi.org/10.1016/j.cor.2021.105271

48. Castellucci PB, Darvish M, Coelho LC (2021) A Benders decomposition algorithm for the time-dependent vehicle routing problem. Technical report CIRRELT-2021-16, Université de Montreal

49. Chen JH, Lee DH, Cao JX (2012) A combinatorial Benders' cuts algorithm for the quayside operation problem at container terminals. Transp Res Part E 48:266–275. https://doi.org/10.1016/j.tre.2011.06.004

50. Chen S, Zen Q, Hu Y (2022) Scheduling optimization for two crossover automated stacking cranes considering relocation. Oper Res: Int J 22:2099–2120. https://doi.org/10.1007/s12351-020-00601-6

51. Chinneck JW (1996) An effective polynomial-time heuristic for the minimum-cardinality IIS set-covering problem. Ann Math Artif Intell 17:127–144. https://doi.org/10.1007/BF02284627

52. Chinneck JW (1997) Finding a useful subset of constraints for analysis in an infeasible linear program. INFORMS J Comput 9:111–129. https://doi.org/10.1287/ijoc.9.2.164

53. Chinneck JW (2001) Fast heuristics for the maximum feasible subsystem problem. INFORMS J Comput 13:171–256. https://doi.org/10.1287/ijoc.13.3.210.12632

54. Chinneck JW, Dravnieks EW (1991) Locating minimal infeasible constraint sets in linear programs. ORSA J Comput 3:157–168. https://doi.org/10.1287/ijoc.3.2.157

55. Chisca DS, Lombardi M, Milano M, O'Sullivan B (2019) Logic-based Benders decomposition for super solutions: an application to the kidney exchange problem. In: Schiex T, de Givry S (eds) Principles and practice of constraint programming (CP2019). Lecture notes in computer science, vol 11802, pp 108–125. https://doi.org/10.1007/978-3-030-30048-7_7

56. Chvátal V (1973) Edmonds polytopes and a hierarchy of combinatorial problems. Discret Math 4:305–337. https://doi.org/10.1016/0012-365X(73)90167-2

57. Ciré AA, Çoban E, Hooker JN (2016) Logic-based Benders decomposition for planning and scheduling: a computational analysis. Knowl Eng Rev 31:440–451. https://doi.org/10.1017/S0269888916000254

58. Çoban E, Heching A, Hooker JN, Scheller-Wolf A (2014) Robust scheduling with logic-based Benders decomposition. In: Lübbecke M, Koster A, Letmangthe P, Madlener R, Peis B, Walther G (eds) Operations research proceedings 2014. Springer, pp 99–105. https://doi.org/10.1007/978-3-319-28697-6_15

59. Çoban E, Hooker JN (2013) Single-facility scheduling by logic-based Benders decomposition. Ann Oper Res 210:245–272. https://doi.org/10.1007/s10479-011-1031-z

60. Codato G, Fischetti M (2006) Combinatorial Benders cuts for mixed-integer linear programming. Oper Res 54:756–766. https://doi.org/10.1287/opre.1060.0286

61. Corréa AI, Langevin A, Rousseau LM (2007) Scheduling and routing of automated guided vehicles: a hybrid approach. Comput Oper Res 34:1688–1707. https://doi.org/10.1016/j.cor.2005.07.004

62. Côté JF, Dell'Amico M, Iori M (2014) Combinatorial Benders cuts for the strip packing problem. Oper Res 62:643–661

63. Côté JF, Gendreau M, Potvin JY (2014) An exact algorithm for the two-dimensional orthogonal packing problem with unloading constraints. Oper Res 62:1126–1141. https://doi.org/10.1287/opre.2014.1307

64. Côté JF, Haouari M, Iori M (2021) Combinatorial Benders decomposition for the two-dimensional bin packing problem. INFORMS J Comput 33:963–978. https://doi.org/10.1287/ijoc.2020.1014

65. Daryalal M, Pouya H, DeSantis MA (2023) Network migration problem: a hybrid logic-based Benders decomposition approach. INFORMS J Comput. https://doi.org/10.1287/ijoc.2023.1280

66. Davies J, Bacchus F (2013) Postponing optimization to speed up MAXSAT solving. In: Schulte C (ed) Principles and practice of constraint programming (CP 2013). Lecture notes in computer science, vol 8124. Springer, pp 247–262

67. Davies J, Bacchus F (2013) Solving MAXSAT by solving a sequence of simpler SAT instances. In: Lee J (ed) Principles and practice of constraint programming (CP 2011). Lecture notes in computer science, vol 6876. Springer, pp 225–239

68. Davies TO, Gange G, Stuckey PJ (2017) Automatic logic-based Benders decomposition with MiniZinc. In: Lübbecke M, Koster A, Letmangthe P, Madlener R, Peis B, Walther G (eds) AAAI conferenece on artificiaal intelligence, pp 787–793

69. Davies TO, Pearce AR, Stuckey P, Lipovetzky N (2015) Sequencing operator counts. In: International conference on automated planning and scheduling (ICAPS). AAAI, pp 61–69. https://doi.org/10.1609/icaps.v25i1.13727

70. Davies TO, Pearce AR, Stuckey P, Lipovetzky N (2016) Sequencing operator counts. In: International joint conference on artificial intelligence (IJCAI), pp 4140–4144

71. Delorme M, Iori M, Martello S (2017) Logic-based Benders' decomposition for orthogonal stock cutting problems. Comput Oper Res 78:290–298. https://doi.org/10.1016/j.cor.2016.09.009

72. Diefenbach H, Emde S, Glock CH, Grosse EH (2022) New solution procedures for the order picker routing problem in U-shaped pick areas with a movable depot. OR Spectrum 44:535–573. https://doi.org/10.1007/s00291-021-00663-8

73. Doi T, Nishui T (2014) Two-level decomposition algorithm for shift scheduling problems. In: IEEE international conference on systems, man and cybernetics, pp 3773–3778. https://doi.org/10.1109/SMC.2014.6974518

74. Elçi Ö, Hooker JN (2023) Strong logic-based Benders cuts for routing and scheduling. Technical report, Carnegie Mellon University

75. Elçi Ö, Hooker JN (2022) Stochastic planning and scheduling with logic-based Benders decomposition. INFORMS J Comput 34:2383–2865. https://doi.org/10.1287/ijoc.2022.1184

76. Emde S (2017) Optimally scheduling interfering and non-interfering cranes. Nav Res Logist 64:476–489. https://doi.org/10.1002/nav.21768

77. Emde S, Zehtabian S, Disser Y (2023) Point-to-point and milk run delivery scheduling: models, complexity results, and algorithms based on Benders decomposition. Ann Oper Res 322:467–496. https://doi.org/10.1007/s10479-022-04891-1

78. Emeretlis A, Theodoridis G, Alefragis P, Voros N (2014) A hybrid ILP-CP model for mapping directed acyclic task graphs to multicore architectures. In: IEEE 28th international parallel and distributed processing symposium. IEEE, pp 176–182. https://doi.org/10.1109/IPDPSW.2014.24

79. Emeretlis A, Theodoridis G, Alefragis P, Voros N (2015) Mapping DAGs on heterogeneous platforms using logic-based Benders decomposition. In: IEEE computer society annual symposium on VLSI, pp 119–124. https://doi.org/10.1109/ISVLSI.2015.98

80. Emeretlis A, Theodoridis G, Alefragis P, Voros N (2017) Static mapping of applications on heterogeneous multi-core platforms combining logic-based Benders decomposition with integer linear programming. ACM Trans Design Autom Electr Syst 23:26:1–26:24. https://doi.org/10.1145/3133219

81. Emeretlis A, Theodoridis G, Alefragis P, Voros N (2022) A multi-stage hybrid approach for mapping applications on heterogeneous multi-core platforms. In: 30th international conference on very large scale integration (VLSI-SoC). IFIP/IEEE, pp 1–6. https://doi.org/10.1109/VLSI-SoC54400.2022.9939643

82. Enayaty-Ahangar F, Rainwater CE, Sharkey TC (2019) A logic-based decomposition approach for multi-period network interdiction models. Omega 87:71–85. https://doi.org/10.1016/j.omega.2018.08.006

83. Erbeyoğlu G, Bilge Ü (2020) A robust disaster preparedness model for effective and fair disaster response. Eur J Oper Res 280:479–494. https://doi.org/10.1016/j.ejor.2019.07.029

84. Erdoğan G, Battarra M, Calvo W (2015) An exact algorithm for the static rebalancing problem arising in bicycle sharing systems. Eur J Oper Res 245:667–679. https://doi.org/10.1016/j.ejor.2015.03.043

85. Esmaeilbeigi R, Mak-Hau V, Yearwood J, Nguyen V (2022) The multiphase course timetabling problem. Eur J Oper Res 300:1098–1119. https://doi.org/10.1016/j.ejor.2021.10.014

86. Fachini RE, Armentano VA (2020) Logic-based Benders decomposition for the heterogeneous fixed fleet vehicle routing problem with time windows. Comput Ind Eng 148. https://doi.org/10.1016/j.cie.2020.106641

87. Fang K, Wang S, Pinedo ML, Chen L, Chu F (2021) A combinatorial Benders decomposition algorithm for parallel machine scheduling with working-time restrictions. Eur J Oper Res 291:128–146. https://doi.org/10.1016/j.ejor.2020.09.037

88. Fazel-Zarandi MM, Beck JC (2009) Solving a location-allocation problem with logic-based Benders decomposition. In: Gent IP (ed) Principles and practice of constraint programming (CP 2009). Lecture notes in computer science, vol 5732. Springer, New York, pp 344–351. https://doi.org/10.1007/978-3-642-04244-7_28

89. Fazel-Zarandi MM, Beck JC (2012) Using logic-based Benders decomposition to solve the capacity- and distance-constrained plant location problem. INFORMS J Comput 24:387–398. https://doi.org/10.1287/ijoc.1110.0458

90. Fazel-Zarandi MM, Berman O, Beck JC (2013) Solving a stochastic facility location/fleet management problem with logic-based Benders decomposition. IIE Trans 45:896–911. https://doi.org/10.1080/0740817X.2012.705452

91. Flener P, Pearson J, Sellmann M (2009) Static and dynamic structural symmetry breaking. Ann Math Artif Intell 57:37–57. https://doi.org/10.1007/s10472-009-9172-3

92. Fontaine P, Minner S (2022) A branch-and-repair method for three-dimensional bin selection and packing in E-commerce. Oper Res 71:273–288. https://doi.org/10.1287/opre.2022.2369

93. Froger A, Gendreau M, Mendoza JE, Pinson E, Rousseau LM (2017) A branch-and-check approach for a wind turbine maintenance scheduling problem. Comput Oper Res 88:117–136. https://doi.org/10.1016/j.cor.2017.07.001

94. Froger A, Jabali O, Mendoza JE, Laport G (2022) The electric vehicle routing problem with capacitated charging stations. Transp Sci 56:460–482. https://doi.org/10.1287/trsc.2021.1111

95. Furugi A (2022) Sequence-dependent time- and cost-oriented assembly line balancing problems: a combinatorial Benders' decomposition approach. Eng Optim 54:170–184. https://doi.org/10.1080/0305215X.2021.1953003

96. Gavanelli M, Milano M, Holland A, O'Sullivan B (2012) What-if analysis through simulation-optimization hybrids. In: 26th European conference on modelling and simulation (ECMS). European council for modeling and simulation, pp 624–630. https://doi.org/10.7148/2012-0624-0630

97. Gedika R, Rainwater C, Nachtmann H, Pohl EA (2016) Analysis of a parallel machine scheduling problem with sequence dependent setup times and job availability intervals. Eur J Oper Res 251:640–650. https://doi.org/10.1016/j.ejor.2015.11.020

98. Gencosman BC, Begen MA (2022) Exact optimization and decomposition approaches for shelf space allocation. Eur J Oper Res 299:432–447. https://doi.org/10.1016/j.ejor.2021.08.047

99. Gendron B, Garroppo RG, Nencioni G, Scutellà MG, Tavanti L (2013) Benders decomposition for a location-design problem in green wireless local area networks. Electr Notes Discrete Math 41:367–374. https://doi.org/10.1016/j.endm.2013.05.114

100. Gendron B, Lucena A, Salles da Cunha A, Simonetti L (2014) Benders decomposition, branch-and-cut, and hybrid algorithms for the minimum connected dominating set problem. INFORMS J Comput 26:645–657. https://doi.org/10.1287/ijoc.2013.0589

101. Gendron B, Scutellà MG, Garroppo RG, Nencioni G, Tavanti L (2016) A branch-and-Benders-cut method for nonlinear power design in green wireless local area networks. Eur J Oper Res 255:151–162. https://doi.org/10.1016/j.endm.2013.05.114

102. Gent IP, Petrie KE, Puget JF (2006) Symmetry in constraint programming. In: Rossi F, van Beek P, Walsh T (eds) Handbook of constraint programming. Elsevier, Amsterdam, pp 329–376

103. Geoffrion AM (1972) Generalized Benders decomposition. J Optim Theory Appl 10:237–260. https://doi.org/10.1007/BF00934810

104. Gleeson J, Ryan J (1990) Identifying minimally infeasible subsystems of inequalities. ORSA J Comput 2:61–63. https://doi.org/10.1287/ijoc.2.1.61

105. Godbersen G, Kolisch R, Schiffer M (2022) Robust charging network planning for metropolitan taxi fleets. arXiv:2209.07305v1

106. Goldwasser A, Schutt A (2018) Optimal torpedo scheduling. J Artif Intell Res 63:955–986. https://doi.org/10.1613/jair.1.11268

107. Gomeza C, Baker JW (2019) An optimization-based decision support framework for coupled pre- and post-earthquake infrastructure risk management. Struct Saf 77:1–9. https://doi.org/10.1016/j.strusafe.2018.10.002

108. Grenouilleau F, Lahrichi N, Rousseau LM (2020) New decomposition methods for home care scheduling with predefined visits. Comput Oper Res 115. https://doi.org/10.1016/j.cor.2019.104855

109. Guo C, Bodur M, Aleman DM, Urbach DR (2021) Logic-based Benders decomposition and binary decision diagram based approaches for stochastic distributed operating room scheduling. INFORMS J Comput 33:1551–1569. https://doi.org/10.1287/ijoc.2020.1036

110. Guo P, He X, Luan Y, Wang Y (2021) Logic-based Benders decomposition for gantry crane scheduling with transferring position constraints in a rail-road container terminal. Eng Optim 53:86–106. https://doi.org/10.1080/0305215X.2019.1699919

111. Hamdi I, Loukil T (2013) Logic-based Benders decomposition to solve the permutation flow-shop scheduling problem with time lags. In: International conference on modeling, simulation and applied optimization (ICMSAO). IEEE, pp 1–7. https://doi.org/10.1109/ICMSAO.2013.6552689

112. Hamdi I, Loukil T (2013) Permutation flowshop problem with time lags scheduling by logic-based Benders decomposition. In: International conference on control, decision and information technologies (CoDIT). IEEE, pp 851–856. https://doi.org/10.1109/CoDIT.2013.6689654

113. Hamdi I, Loukil T (2015) Upper and lower bounds for the permutation flowshop scheduling problem with minimal time lags. Optim Lett 9:465–482. https://doi.org/10.1007/s11590-014-0761-7

114. Han J, Zhang J, Zeng B, Mao M (2021) Optimizing dynamic facility location-allocation for agricultural machinery maintenance using Benders decomposition. Omega 105. https://doi.org/10.1016/j.omega.2021.102498

115. Harjunkoski I, Grossmann IE (2002) Decomposition techniques for multistage scheduling problems using mixed-integer and constraint programming methods. Comput Chem Eng 26:1533–1552. https://doi.org/10.1016/S0098-1354(02)00100-X

116. Harris MG, Forbes MA, Taimre T (2022) Logic-based Benders decomposition for wild fire suppression. arXiv:2209.01371v1 . https://doi.org/10.48550/arXiv.2209.01371

117. Heching A, Hooker JN (2016) Scheduling home hospice care with logic-based Benders decomposition. In: Quimper CG (ed) CPAIOR 2016 proceedings, pp 187–197. https://doi.org/10.1007/978-3-319-33954-2_14

118. Heching A, Hooker JN, Kimura R (2019) A logic-based Benders approach to home healthcare delivery. Transp Sci 53:510–522. https://doi.org/10.1287/trsc.2018.0830

119. Hedman KW, Oren SS, O'Neill RP (2011) Revenue adequacy constrained optimal transmission switching. In: 44th Hawaii international conference on system sciences. IEEE. https://doi.org/10.1109/HICSS.2011.361

120. Hladik PE, Cambazard H, Déplanche AM, Jussien N (2008) Solving a real-time allocation problem with constraint programming. J Syst Softw 81:132–149. https://doi.org/10.1016/j.jss.2007.02.032

121. Ho MH, Hnaien F, Dugardin F (2022) Exact method to optimize the total electricity cost in two-machine permutation flow shop scheduling problem under time-of-use tariff. Comput Oper Res 144. https://doi.org/10.1016/j.cor.2022.105788

122. Hooker JN (2000) Logic-based methods for optimization: combining optimization and constraint satisfaction. Wiley, New York

123. Hooker JN (2007) Planning and scheduling by logic-based Benders decomposition. Oper Res 55:588–602. https://doi.org/10.1287/opre.1060.0371

124. Hooker JN (2012) Integrated methods for optimization, 2nd edn. Springer

125. Hooker JN (2019) Logic-based Benders decomposition for large-scale optimization. In: Velásquez-Bermúdez J, Khakifirooz M, Fathi M (eds) Large scale optimization in supply chains and smart manufacturing. Springer optimization and its applications, vol 149. Springer, Berlin, pp 1–26. https://doi.org/10.1007/978-3-030-22788-3_1

126. Hooker JN, Ottosson G (2003) Logic-based Benders decomposition. Math Program 96:33–60. https://doi.org/10.1007/s10107-003-0375-9

127. Hooker JN, Yan H (1995) Verifying logic circuits by Benders decomposition. In: Saraswat V, Hentenryck PV (eds) Principles and practice of constraint programming: the newport papers. MIT Press, Cambridge, MA, pp 267–288

128. Hooshmand F, Amerehi F, MirHassani SA (2020) Logic-based Benders decomposition algorithm for contamination detection problem in water networks. Comput Oper Res 115. https://doi.org/10.1016/j.cor.2019.104840

129. Horváth M, Kis T, Kovács A, Fekula M (2022) Assembly planning by disjunctive programming and geometrical reasoning. Comput Oper Res 138. https://doi.org/10.1016/j.cor.2021.105603

130. Hosseininasab A, van Hoeve WJ (2021) Exact multiple sequence alignment by synchronized decision diagrams. INFORMS J Comput 33:721–738. https://doi.org/10.1287/ijoc.2019.0937

131. Hu J, Mitchell JE, Pang JS (2012) An LPCC approach to nonconvex quadratic programs. Math Program 133:243–277. https://doi.org/10.1007/s10107-010-0426-y

132. Hu J, Mitchell JE, Pang JS, Bennett KP, Kunapuli G (2008) On the global solution of linear programs with linear complementarity constraints. SIAM J Optim 19:445–471. https://doi.org/10.1137/07068463x

133. Hua Q, Zhang Z, Baldacci R, Tarantilis CD, Zachariadis E (2022) The bus sightseeing problem. Int Trans Oper Res. https://doi.org/10.1111/itor.13160

134. Huang X, Zhang B, Li C (2021) Platform profit maximization on service provisioning in mobile edge computing. IEEE Trans Veh Technol 70. https://doi.org/10.1109/TVT.2021.3124483

135. Hussain T, Frey G (2008) Solving the deployment problem of IEC 61499 applications. In: 17th world congress of the international federation of automatic control (IFAC). IFAC, pp 8321–8326. https://doi.org/10.3182/20080706-5-KR-1001.1781

136. Jain V, Grossmann IE (2001) Algorithms for hybrid MILP/CP models for a class of optimization problems. INFORMS J Comput 13:258–276. https://doi.org/10.1287/ijoc.13.4.258.9733

137. Jara-Moroni F, Mitchell JE, Pang JS, Wachter A (2020) An enhanced logical Benders approach for linear programs with complementarity constraints. J Global Optim 77. https://doi.org/10.1007/s10898-020-00905-z

138. Jeroslow RE, Wang J (1990) Solving propositional satisfiability problems. Ann Math Artif Intell 1:167–188. https://doi.org/10.1007/BF01531077

139. Jiang SL, Xu C, Zhang L, Ma Y (2023) A decomposition-based two-stage online scheduling approach and its integrated system in the hybrid flow shop of steel industry. Expert Syst Appl. https://doi.org/10.1016/j.eswa.2022.119200

140. Jones E (2020) Decomposing systems: illustrating the utility of distributed energy resources with decomposition techniques. In: Cromarty L, Shirwaiker R, Wang P (eds) IIE annual conference. Norcross, pp 477–482

141. Jungwirth A, Desaulniers G, Frey M, Kolische R (2022) Exact branch-price-and-cut for a hospital therapist scheduling problem with flexible service locations and time-dependent location capacity. INFORMS J Comput 34:1157–1175. https://doi.org/10.1287/ijoc.2021.1119

142. Junker U (2001) QuickXplain: conflict detection for arbitrary constraint propagation algorithms. In: IJCAI01 workshop on modeling and solving problems with constraints (CONS-1). Seattle, USA

143. Juvin C, Houssin L, Lopez P (2023) Logic-based Benders decomposition for the preemptive flexible job-shop scheduling problem. Comput Oper Res 152. https://doi.org/10.1016/j.cor.2023.106156

144. Kafle B, Gange G, Schachte P, Søndergaard H, Stuckey PJ (2017) A Benders decomposition approach to deciding modular linear integer arithmetic. In: Gaspers S, Walsh T (eds) International conference on theory and applications of satisfiability testing, pp 380–397. https://doi.org/10.1007/978-3-319-66263-3_24

145. Kaizer WL, Pereira AG, Ritt M (2020) Sequencing operator counts with state-space search. In: International conference on automated planning and scheduling (ICAPS). AAAI, pp 166–174. https://doi.org/10.1609/icaps.v30i1.6658

146. Kamran MA, Karimi B, Dellaert N, Demeulemeester E (2019) Adaptive operating rooms planning and scheduling: a rolling horizon approach. Oper Res Health Care 22. https://doi.org/10.1016/j.orhc.2019.100200

147. Kang M, Lee C (2021) An exact algorithm for heterogeneous drone-truck routing problem. Transp Sci 55:1088–1112. https://doi.org/10.1287/trsc.2021.1055

148. Karlsson E, Rönnberg E (2021) Strengthening of feasibility cuts in logic-based Benders decomposition. In: Stuckey PJ (ed) CPAIOR 2021 proceedings. Lecture notes in computer science, vol 12735. Springer, pp 45–61. https://doi.org/10.1007/978-3-030-78230-6_3

149. Karlsson E, Rönnberg E (2022) Logic-based Benders decomposition with a partial assignment acceleration technique for avionics scheduling. Comput Oper Res 146. https://doi.org/10.1016/j.cor.2022.105916

150. Kinable J, Trick M (2014) A logic-based Benders approach to the concrete delivery problem. In: Simonis H (ed) CPAIOR 2014 proceedings. Lecture notes in computer science, vol 8451. Springer, pp 176–192. https://doi.org/10.1007/978-3-319-07046-9_13

151. Kloimüllner C, Papazek P, Hu B, Raidl GR (2015) A cluster-first route-second approach for balancing bicycle sharing systems. In: International conference on computer aided systems theory (EUROCAST). Lecture notes in computer science, vol 9520. Springer, pp 439–446. https://doi.org/10.1007/978-3-319-27340-2_55

152. Kloimüllner C, Raidl GR (2017) Full-load route planning for balancing bike sharing systems by logic-based Benders decomposition. Networks 69:439–446. https://doi.org/10.1002/net.21736

153. Kress D, Müller D, Nossack J (2019) A worker constrained flexible job shop scheduling problem with sequence-dependent setup times. OR Spectrum 41:179–217. https://doi.org/10.1007/s00291-018-0537-z

154. Lam E, Gange G, Stuckey PJ, Van Hentenryck P, Dekker JJ (2020) Nutmeg: a MIP and CP hybrid solver using branch-and-check. SN Oper Res Forum 1. https://doi.org/10.1007/s43069-020-00023-2

155. Lam E, Van Hentenryck P (2017) Branch-and-check with explanations for the vehicle routing problem with time windows. In: Beck JC (ed) Principles and practice of constraint programming (CP 2017). Lecture notes in computer science, vol 10416. Springer, pp 579–595. https://doi.org/10.1007/978-3-319-66158-2

156. Lamorgese L, Mannino C (2015) An exact decomposition approach for the real-time train dispatching problem. Oper Res 63:48–64. https://doi.org/10.1287/opre.2014.1327

157. Lamorgese L, Mannino C (2019) A noncompact formulation for job-shop scheduling problems in traffic management. Oper Res 19:1586–1609. https://doi.org/10.1287/opre.2018.1837

158. Lamorgese L, Mannino C, Piacentini M (2016) Optimal train dispatching by Benders'-like reformulation. Transp Sci 60:910–925. https://doi.org/10.1287/trsc.2015.0605

159. Laporte G, Louveaux FV (1993) The integer L-shaped method for stochastic integer programs with complete recourse. Oper Res Lett 13:133–142. https://doi.org/10.1016/0167-6377(93)90002-X

160. Lee C, Cho D, Park S (2019) A combinatorial Benders decomposition algorithm for the directed multiflow network diversion problem. Mil Oper Res 24:23–40. https://doi.org/10.2307/26609633

161. Leutwiler F, Corman F (2022) A logic-based Benders decomposition for microscopic railway timetable planning. Eur J Oper Res 303:525–540. https://doi.org/10.1016/j.ejor.2022.02.043

162. Li S, Chen W, Chen Y, Chen C, Zheng Z (2019) Makespan-minimized computation offloading for smart toys in edge-cloud computing. Electr Commerce Res Appl 37. https://doi.org/10.1016/j.elerap.2019.100884

163. Li S, Negenborn RR, Lodewijks G (2016) A logic-based Benders decomposition approach to improve coordination of inland vessels for inter-terminal transport. In: International conference on computational logistics (ICCL). Lecture notes in computer science, vol 9855. Springer, pp 96–115. https://doi.org/10.1007/978-3-319-44896-1_7

164. Li S, Negenborn RR, Lodewijks G (2017) Closed-loop coordination of inland vessels operations in large seaports using hybrid logic-based Benders decomposition. Transp Res Part E 97:1–21. https://doi.org/10.1016/j.tre.2016.10.013

165. Li X, Aneja YP (2020) A branch-and-Benders-cut approach for the fault tolerant regenerator location problem. Comput Oper Res 115. https://doi.org/10.1016/j.cor.2019.104847

166. Li Y, Côté JF, Coelho LC, Zhang C, Zhang S (2023) Order assignment and scheduling under processing and distribution time uncertainty. Eur J Oper Res 305:148–163. https://doi.org/10.1016/j.ejor.2022.05.033

167. Li Y, Li Y, Cheng J, Wu P (2022) Order assignment and scheduling for personal protective equipment production during the outbreak of epidemics. IEEE Trans Autom Sci Eng 19:692–708. https://doi.org/10.1109/TASE.2021.3137025

168. Li Y, Wen X, Choi TM, Chung SH (2022) Optimal establishments of massive testing programs to combat COVID-19: a perspective of parallel-machine scheduling-location (ScheLoc) problem. IEEE Trans Eng Manage. https://doi.org/10.1109/TEM.2022.3199039

169. Li Y, Wu P (2022) Novel formulations and logic-based Benders decomposition for the integrated parallel machine scheduling and location problem. INFORMS J Comput 34:1048–1069. https://doi.org/10.1287/ijoc.2021.1113

170. van Lieshout RN, Bouman PC, Huisman D (2020) Determining and evaluating alternative line plans in out-of-control situations. Transp Sci 54:740–761. https://doi.org/10.1287/trsc.2019. 0945

171. Lindh E, Olsson K, Rönnberg E (2022) Scheduling of an underground mine by combining logic-based Benders decomposition and a priority-based heuristic. In: De Causmaecker P, Özcan E, Vanden Berghe G (eds) Proceedings of the 13th international conference on the practice and theory of automated timetabling (PATAT 2022), vol 3, pp 95–114

172. van Loon JNM (1981) Irreducibly inconsistent systems of linear inequalities. Eur J Oper Res 8:283–288. https://doi.org/10.1016/0377-2217(81)90177-6

173. Mancini S, Ciavotta M, Meloni C (2021) The multiple multidimensional knapsack with family-split penalties. Eur J Oper Res 289:987–998. https://doi.org/10.1016/j.ejor.2019.07.052

174. Mancini S, Gansterer M (2021) Vehicle scheduling for rental-with-driver services. Trans Res Part E 156. https://doi.org/10.1016/j.tre.2021.102530

175. Maravelias CT (2006) A decomposition framework for the scheduling of single- and multi-stage processes. Comput Chem Eng 30:407–420. https://doi.org/10.1016/j.compchemeng.2005.09. 011

176. Maravelias CT, Grossmann IE (2004) A hybrid MILP/CP decomposition approach for the continuous time scheduling of multipurpose batch plants. Comput Chem Eng 28:1921–1949. https://doi.org/10.1016/j.compchemeng.2004.03.016

177. Maravelias CT, Grossmann IE (2004) Using MILP and CP for the scheduling of batch chemical processes. In: Régin JC, Rueher M(eds) CPAIOR 2004 proceedings. Lecture notes in computer science, vol 3011. Springer, pp 1–20. https://doi.org/10.1007/978-3-540-24664-0_1

178. Margot F (2010) Symmetry in integer linear programming. In: Jünger M, Liebling TM, Naddef D, Nemhauser GL, Pulleyblank WR, Reinelt G, Rinaldi G, Wolsey LA (eds) 50 years of integer programming 1958–2008. Springer, New York, pp 647–686

179. Marques Silva JP, Sakallah KA (1996) GRASP–a new search algorithm for satisfiability. In: International conference on computer-aided design. IEEE, pp 220–227. https://doi.org/10.1109/ ICCAD.1996.569607

180. Martínez KP, Adulyasak Y, Jans R (2022) Logic-based Benders decomposition for integrated process configuration and production planning problems. INFORMS J Comput 34. https://doi. org/10.1287/ijoc.2021.1079

181. Martínez KP, Adulyasak Y, Jans R, Morabito R, Toso EAV (2019) An exact optimization approach for an integrated process configuration, lot-sizing, and scheduling problem. Comput Oper Res 103:310–323. https://doi.org/10.1016/j.cor.2018.10.005

182. Maschler J, Raidl G (2017) A logic-based Benders decomposition approach for the 3-staged strip packing problem. In: Dörner KF, Ljubic I, Pflug G, Tragler G (eds) Operations research proceedings 2015. Springer, pp 393–399. https://doi.org/10.1007/978-3-319-42902-1_53

183. Michels AS, Lopes TC, Sikora CGS, Magatão L (2019) A Benders' decomposition algorithm with combinatorial cuts for the multi-manned assembly line balancing problem. Eur J Oper Res 278:796–808. https://doi.org/10.1016/j.ejor.2019.05.001

184. Mistry M, D'Iddio AC, Huth M, Misener R (2018) Satisfiability modulo theories for process systems engineering. Comput Chem Eng 113:98–114. https://doi.org/10.1016/j.compchemeng. 2018.03.004

185. Módos I, Šůcha P, Hanzálek Z (2017) Algorithms for robust production scheduling with energy consumption limits. Comput Ind Eng 112:391–408. https://doi.org/10.1016/j.cie.2017.08.011

186. Mohamed IB, Klibi W, Sadykov R, Şen H, Vanderbeck F (2023) The two-echelon stochastic multi-period capacitated location-routing problem. Eur J Oper Res 306:645–667. https://doi. org/10.1016/j.ejor.2022.07.022

187. Monemi RN, Gelareh S (2017) Network design, fleet deployment and empty repositioning in liner shipping. Transp Res Part E 108:60–79. https://doi.org/10.1016/j.tre.2017.07.005

188. Moradi S, MirHassani SA, Hooshmand F (2019) Efficient decomposition-based algorithm to solve long-term pipeline scheduling problem. Pet Sci 16:1159–1175. https://doi.org/10.1007/s12182-019-00359-3

189. Mubarak M, Üster H, Abdelghany K, Khodayar M (2021) Strategic network design and analysis for in-motion wireless charging of electric vehicles. Trans Res Part E 145. https://doi.org/10.1016/j.tre.2020.102179

190. Naderi B, Begen MA, Zaric GS (2022) Type-2 integrated process-planning and scheduling problem: reformulation and solution algorithms. Comput Oper Res 142. https://doi.org/10.1016/j.cor.2022.105728

191. Naderi B, Begen MA, Zaric GS, Roshanaei V (2023) A novel and efficient exact technique for integrated staffing, assignment, routing, and scheduling of home care services under uncertainty. Omega 116. https://doi.org/10.1016/j.omega.2022.102805

192. Naderi B, Govindan K, Soleiman H (2020) A Benders decomposition approach for a real case supply chain network design with capacity acquisition and transporter planning: wheat distribution network. Ann Oper Res 291:685–705. https://doi.org/10.1007/s10479-019-03137-x

193. Naderi B, Roshanaei V (2020) Branch-relax-and-check: a tractable decomposition method for order acceptance and identical parallel machine scheduling. Eur J Oper Res 291:811–827. https://doi.org/10.1016/j.ejor.2019.10.014

194. Naderi B, Roshanaei V (2022) Critical-path-search logic-based Benders decomposition approaches for flexible job shop scheduling. INFORMS J Optim 4. https://doi.org/10.1287/ijoo.2021.0056

195. Naderi B, Roshanaei V, Begen MA, Aleman DM, Urbach DR (2021) Increased surgical capacity without additional resources: generalized operating room planning and scheduling. Prod Oper Manag 30:2608–2635. https://doi.org/10.1111/poms.13397

196. Naoum-Sawaya J, Elhedhli S (2010) A nested Benders decomposition approach for telecommunication network planning. Nav Res Logist 57:519–539. https://doi.org/10.1002/nav.20419

197. Nemhauser GL, Wolsey LA (1999) Integer and combinatorial optimization. Wiley, New York. https://doi.org/10.1002/9781118627372

198. Nishi T, Hiranaka Y, Grossmann IE (2011) A bilevel decomposition algorithm for simultaneous production scheduling and conflict-free routing for automated guided vehicles. Comput Oper Res 38:876–888. https://doi.org/10.1016/j.cor.2010.08.012

199. Nishi T, Sugiyama T, Inuiguchi M (2014) Two-level decomposition algorithm for crew rostering problems with fair working condition. Eur J Oper Res 237:465–473. https://doi.org/10.1016/j.ejor.2014.02.010

200. Nossack J, Briskorn D, Pesch E (2018) Container dispatching and conflict-free yard crane routing in an automated container terminal. Transp Sci 52:1059–1076. https://doi.org/10.1287/trsc.2017.0811

201. Nurdiansyah R, Hong I (2018) Combinatorial Benders' cut for the admission control decision in flow shop scheduling problems with queue time constraints. In: Moon I, Lee G, Park J, Kiritsis D, von Cieminski G (eds) Advances in production management systems: production management for data-driven, intelligent, collaborative, and sustainable manufacturing (APMS). Springer, pp 399–405. https://doi.org/10.1007/978-3-319-99704-9_49

202. Osman H, Demirli K (2010) A bilinear goal programming model and a modified Benders decomposition algorithm for supply chain reconfiguration and supplier selection. Int J Prod Econ 124:97–105. https://doi.org/10.1016/j.ijpe.2009.10.012

203. Osman H, Demirli K (2012) Integrated safety stock optimization for multiple sourced stock-points facing variable demand and lead time. Int J Prod Econ 135:299–307. https://doi.org/10.1016/j.ijpe.2011.08.004

204. Paradiso R, Georghiou A, Dabia S, Tönissen D (2022) Exact and approximate schemes for robust optimization problems with decision dependent information discovery. arXiv:2208.04115v1

205. Polyakovskiy S, M'Hallah R (2021) Just-in-time two-dimensional bin packing. Omega 102. https://doi.org/10.1016/j.omega.2020.10231

206. Prékopa A (2013) Stochastic programming. Springer, New York. https://doi.org/10.1007/978-94-017-3087-7

207. Pugliese L, Ferone D, Macrina G, Festa P, Guerriero F (2023) The crowd-shipping with penalty cost function and uncertain travel times. Omega 115. https://doi.org/10.1016/j.omega.2022.102776

208. Pugliese L, Guerriero F, Scutellà MG (2021) The last-mile delivery process with trucks and drones under uncertain energy consumption. J Optim Theory Appl 191:31–67. https://doi.org/10.1007/s10957-021-01918-8

209. Qin T, Du Y, Sha M (2016) Evaluating the solution performance of IP and CP for berth allocation with time-varying water depth. Transp Res Part E 87:167–185. https://doi.org/10.1016/j.tre.2016.01.007

210. Rahmaniani R, Crainic TG, Gendreau M, Rei W (2017) The Benders decomposition algorithm: a literature review. Eur J Oper Res 259:801–817. https://doi.org/10.1016/j.ejor.2016.12.005

211. Rasmussen RV (2008) Scheduling a triple round robin tournament for the best Danish soccer league. Eur J Oper Res 185:795–810. https://doi.org/10.1016/j.ejor.2006.12.050

212. Rasmussen RV, Trick MA (2007) A Benders approach to the constrained minimum break problem. Eur J Oper Res 177:198–213. https://doi.org/10.1016/j.ejor.2005.10.063

213. Regue R, Masoud N, Recker W (2016) Car2work: shared mobility concept to connect commuters with workplaces. Transp Res Rec: J Transp Res Board 2542:102–110. https://doi.org/10.3141/2542-12

214. Restrepo MI, Gendron B, Rousseau LM (2018) Combining Benders decomposition and column generation for multi-activity tour scheduling. Comput Oper Res 93:151–165. https://doi.org/10.1016/j.cor.2018.01.014

215. Riazi S, Diding T, Falkman P, Bengtsson K, Lennartson B (2019) Scheduling and routing of AGV for large-scale flexible manufacturing systems. In: 15th international conference on automation science and engineering (CASE). IEEE, pp 891–896. https://doi.org/10.1109/COASE.2019.8842849

216. Riazi S, Lennartson B (2021) Using CP/SMT solvers for scheduling and routing of AGVs. IEEE Trans Autom Sci Eng 18:218–229. https://doi.org/10.1109/TASE.2020.3012879

217. Riedler M, Raidl GR (2018) Solving a selective dial-a-ride problem with logic-based Benders decomposition. Comput Oper Res 96:30–54. https://doi.org/10.1016/j.cor.2018.03.008

218. Riise A, Mannino C, Lamorgese L (2017) Recursive logic-based Benders' decomposition for multi-mode outpatient scheduling. Eur J Oper Res 257:439–455. https://doi.org/10.1016/j.ejor.2016.06.015

219. Rist Y, Forbes M (2022) A column generation and combinatorial Benders decomposition algorithm for the selective dial-a-ride-problem. Comput Oper Res 140. https://doi.org/10.1016/j.cor.2021.105649

220. Roshanaei V, Booth KEC, Aleman DM, Urbach DR, Beck JC (2020) Branch-and-check methods for multi-level operating room planning and scheduling. Inter J Prod Econ 220. https://doi.org/10.1016/j.ijpe.2019.07.006

221. Roshanaei V, Luong C, Aleman DM, Urbach D (2017) Collaborative operating room planning and scheduling. INFORMS J Comput 29:558–580. https://doi.org/10.1287/ijoc.2017.0745

222. Roshanaei V, Luong C, Aleman DM, Urbach D (2017) Propagating logic-based Benders decomposition approaches for distributed operating room scheduling. Eur J Oper Res 257:439–455. https://doi.org/10.1016/j.ejor.2016.08.024

223. Roshanaei V, Luong C, Aleman DM, Urbach D (2020) Reformulation, linearization, and decomposition techniques for balanced distributed operating room scheduling. Omega 93. https://doi.org/10.1016/j.omega.2019.03.001

224. Roshanaei V, Naderi B (2021) Solving integrated operating room planning and scheduling: logic-based Bender decomposition versus branch-price-and-cut. Eur J Oper Res 293:65–78. https://doi.org/10.1016/j.ejor.2020.12.004

225. Sadykov R (2008) A branch-and-check algorithm for minimizing the weighted number of late jobs on a single machine with release dates. Eur J Oper Res 189:1283–1304

226. Şahin M, Kellegöz T (2022) Benders' decomposition based exact solution method for multi-manned assembly line balancing problem with walking workers. Ann Oper Res 321:507–540. https://doi.org/10.1007/s10479-022-05118-z

227. Sapucaia A, Ciré AA, de Rezende PJ, de Rezende SF, de Souza CC (2023) Covering points with unit disks under color constraints. Technical report, University of Campinas, University of Toronto and Lund University. https://doi.org/10.2139/ssrn.4334380

228. Sekhavatmanesh H, Cherkaoui R (2020) A novel decomposition solution approach for the restoration problem in distribution networks. IEEE Trans Power Syst 35:3810–3824. https://doi.org/10.1109/TPWRS.2020.2982502

229. Shapiro A, Dentcheva D, Ruszeczyński A (2009) Lectures on stochastic programming. SIAM, Philadelphia

230. Sikora CGS (2021) Benders' decomposition for the balancing of assembly lines with stochastic demand. Eur J Oper Res 202:108–124. https://doi.org/10.1016/j.ejor.2020.10.019

231. Silva A, Coelho LC, Darvish M, Rei W (2022) Robotic mobile fulfillment system with pod repositioning for energy saving. Technical report 2022-12, CIRRELT

232. Sorkhoh I, Assi C, Ebrahimi D, Sharafeddine S (2022) Optimizing information freshness for MEC-enabled cooperative autonomous driving. IEEE Trans Intell Transp Syst 23:13127–13140. https://doi.org/10.1109/TITS.2021.3119961

233. Stuckey PJ (2023) Advanced modeling for discrete optimization: square packing. https://www.coursera.org/lecture/advanced-modeling/2-4-1-square-packing-oXHGs

234. Sun D, Meng Y, Tang L, Liu J, Huang B, Yang J (2020) Storage space allocation problem at inland bulk material stockyard. Trans Res Part E 134. https://doi.org/10.1016/j.tre.2020.101856

235. Sun D, Tang L (2013) Benders approach for the raw material transportation scheduling problem in steel industry. In: 10th international conference on control and automation (ICCA). IEEE, pp 481–484. https://doi.org/10.1109/ICCA.2013.6565191

236. Sun D, Tang L, Baldacci R (2019) A Benders decomposition-based framework for solving quay crane scheduling problems. Eur J Oper Res 273:504–515. https://doi.org/10.1016/j.ejor.2018.08.009

237. Sun D, Tang L, Baldacci R, Lim A (2021) An exact algorithm for the unidirectional quay crane scheduling problem with vessel stability. Eur J Oper Res 291:271–283. https://doi.org/10.1016/j.ejor.2020.09.033

238. Taşkın ZC, Cevik M (2013) Combinatorial Benders cuts for decomposing IMRT fluence maps using rectangular apertures. Comput Oper Res 40:2178–2186. https://doi.org/10.1016/j.cor.2011.07.005

239. Taşkın ZC, Smith JC, Ahmed S, Schaefer AJ (2009) Cutting plane algorithms for solving a stochastic edge-partition problem. Discret Optim 6:420–435

240. Tan Y, Terekhov D (2018) Logic-based Benders decomposition for two-stage flexible flow shop scheduling with unrelated parallel machines. In: Bagheri E, Cheung J (eds) Canadian AI proceedings. Springer, pp 60–781. https://doi.org/10.1007/978-3-319-89656-4_5

241. Tanga L, Sun D, Liu J (2016) Integrated storage space allocation and ship scheduling problem in bulk cargo terminals. IIE Trans 48:428–439. https://doi.org/10.1080/0740817X.2015.1063791

242. Terekhov D, Beck JC, Brown KN (2007) Solving a stochastic queueing design and control problem with constraint programming. In: Proceedings of the 22nd national conference on artificial intelligence (AAAI 2007), vol 1. AAAI Press, pp 261–266

243. Terekhov D, Beck JC, Brown KN (2009) A constraint programming approach for solving a queueing design and control problem. INFORMS J Comput 21:549–561. https://doi.org/10.1287/ijoc.1080.0307

244. Thorsteinsson E (2001) Branch and check: a hybrid framework integrating mixed integer programming and constraint logic programming. In: Walsh T (ed) Principles and practice of constraint programming (CP 2001). Lecture notes in computer science, vol 2239. Springer, pp 16–30. https://doi.org/10.1007/3-540-45578-7_2

245. Timpe C (2002) Solving planning and scheduling problems with combined integer and constraint programming. OR Spectrum 24:431–448. https://doi.org/10.1007/s00291-002-0107-1

246. Tran T, Araujo A, Beck JC (2016) Decomposition methods for the parallel machine scheduling problem with setups. INFORMS J Comput 28:83–95. https://doi.org/10.1287/ijoc.2015.0666

247. Tran TT, Beck JC (2012) Logic-based Benders decomposition for alternative resource scheduling with sequence dependent setups. In: European conference on artificial intelligence (ECAI). Frontiers in artificial intelligence and applications, vol 242. IOS Press, pp 774–779

248. Trick M, Yıldız H (2007) Benders cuts guided large neighborhood search for the traveling umpire problem. In: Hentenryck PV, Wolsey L (eds) CPAIOR proceedings. Lecture notes in computer science, vol 4510. Springer, pp 332–345. https://doi.org/10.1007/978-3-540-72397-4_24

249. Trick M, Yıldız H (2011) Benders cuts guided large neighborhood search for the traveling umpire problem. Naval Res Log 771–781. https://doi.org/10.1002/nav.20482

250. Ünsal Ö (2021) An extended formulation of moldable task scheduling problem and its application to quay crane assignments. Expert Syst Appl 185. https://doi.org/10.1016/j.eswa.2021.115617

251. Ünsal Ö, Oğuz C (2019) An exact algorithm for integrated planning of operations in dry bulk terminals. Trans Res Part E 126. https://doi.org/10.1016/j.tre.2019.03.018

252. Van Bulck D, Goossens D (2022) Optimizing rest times and differences in games played: an iterative two-phase approach. J Sched 25:261–271. https://doi.org/10.1007/s10951-021-00717-3

253. Verstichel J, Kinable J, De Causmaecker P, Vanden Berghe G (2015) A combinatorial Benders' decomposition for the lock scheduling problem. Comput Oper Res 54:117–128. https://doi.org/10.1016/j.cor.2014.09.00

254. Wang S, Wu R, Chuy F, Yuz J (2022) Unrelated parallel machine scheduling problem with special controllable processing times and setups. Comput Oper Res 148. https://doi.org/10.1016/j.cor.2022.105990

255. Wang S, Wu R, Chuy F, Yuz J (2023) An exact decomposition method for unrelated parallel machine scheduling with order acceptance and setup times. Comput Ind Eng 175. https://doi.org/10.1016/j.cie.2022.108899

256. Warwicker JA, Rebennack S (2022) Generating optimal robust continuous piecewise linear regression with outliers through combinatorial Benders decomposition. IISE Trans. https://doi.org/10.1080/24725854.2022.2107249

257. Wheatley D, Gzara F, Jewkes E (2015) Logic-based Benders decomposition for an inventory-location problem with service constraints. Omega 55:10–23. https://doi.org/10.1016/j.omega.2015.02.001

258. Wu X, Guo P, Wang Y, Wang Y (2022) Decomposition approaches for parallel machine scheduling of step-deteriorating jobs to minimize total tardiness and energy consumption. Complex Intell Syst 8:1339–1354. https://doi.org/10.1007/s40747-021-00601-9

259. Xia Q, Eremin A, Wallace M (2004) Problem decomposition for traffic diversions. In: Régin JC, Rueher M (eds) CPAIOR 2004 proceedings. Lecture notes in computer science, vol 3011. Springer, pp 348–363. https://doi.org/10.1007/978-3-540-24664-0_24

260. Xue Z, Zhang C, Yang P, Miao L (2015) A combinatorial Benders' cuts algorithm for the local container drayage problem. Math Probl Eng (Hindawi). https://doi.org/10.1155/2015/134763

261. Yu Q, Adulyasak Y, Rousseau LM, Zhu N, Ma S (2022) Team orienteering with time-varying profit. INFORMS J Comput 34:262–280. https://doi.org/10.1287/ijoc.2020.1026

262. Yunes TH, Aron I, Hooker JN (2010) An integrated solver for optimization problems. Oper Res 58:342–356. https://doi.org/10.1287/opre.1090.0733

263. Zhang C, Li Y, Cao J, Wen X (2022) On the mass COVID-19 vaccination scheduling problem. Comput Oper Res 142. https://doi.org/10.1016/j.cor.2022.105704

264. Zhang M, Zheng N, Li H, Gu Z (2018) A decomposition-based approach to optimization of TTP-based distributed embedded systems. J Syst Architect 91:53–61. https://doi.org/10.1016/j.sysarc.2018.07.006

265. Zhang Q, Wang Z, Huang M, Yu Y, Fang SC (2022) Heterogeneous multi-depot collaborative vehicle routing problem. Transp Res Part B 160. https://doi.org/10.1016/j.trb.2022.03.004

266. Zhang Y, Lin WH, Huang M, Hu X (2021) Multi-warehouse package consolidation for split orders in online retailing. Eur J Oper Res 289:1040–1055. https://doi.org/10.1016/j.ejor.2019.07.00

267. Zhang Z, Denton BT, Morgan TM (2022) Optimization of active surveillance strategies for heterogeneous patients with prostate cancer. Prod Oper Manag 31:4021–4037. https://doi.org/10.1111/poms.13800

268. Zhang Z, Song X, Huang H, Zhou X, Yin Y (2022) Logic-based Benders decomposition method for the seru scheduling problem with sequence-dependent setup time and DeJong's learning effect. Eur J Oper Res 297:866–877. https://doi.org/10.1016/j.ejor.2021.06.017

269. Zohali H, Naderi B, Roshanaei V (2022) Solving the type-2 assembly line balancing with setups using logic-based Benders decomposition. INFORMS J Comput 34:315–332. https://doi.org/10.1287/ijoc.2020.1015